SCIENCE, TECHNOLOGY, & SOCIETY

POPULATIONS

PROJECT CONSULTANTS

JON L. HARKNESS
WAUSAU, WISCONSIN, PUBLIC SCHOOLS

DAVID M. HELGREN
SAN JOSE STATE UNIVERSITY

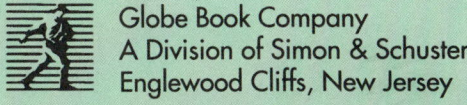
Globe Book Company
A Division of Simon & Schuster
Englewood Cliffs, New Jersey

Project Consultants

Jon L. Harkness is the Science Coordinator and a teacher for the Wausau School district in Wausau, Wisconsin. He is also a founding member and on the board of directors of the National Association for Science, Technology, and Society.

David M. Helgren teaches in the Department of Geology and Environmental Studies at San Jose State University. He has authored both elementary and high school level social studies textbooks.

Executive Editor: Karen Kennedy Gotimer
Editors: Laura Sprankel Baselice, Thomas M. Frado, Robert J. Hope, Kim Merlino, Natania Mlawer, Stedman Willard
Art Director: Nancy Sharkey
Text Design: Siren Design
Cover Design: B B & K Designs, Inc.
Cover Photo: Bill Hickey/Image Bank
Photo Researcher: Jenifer Hixson
Production Manager: Lisa Cowart

Acknowledgments can be found beginning on page 268.

Copyright © 1993 by Globe Book Company, 190 Sylvan Avenue, Englewood Cliffs, New Jersey 07632. All rights reserved. No part of this book may be kept in an information storage or retrieval system, transmitted or reproduced in any form or by any means without the prior written permission of the publisher.

Printed in the United States of America 1 2 3 4 5 6 7 8 9 10 96 95 94 93 92

ISBN 0-8359-0456-3

GLOBE BOOK COMPANY
A Division of Simon & Schuster
Englewood Cliffs, New Jersey

This book is printed on recycled paper.

ACKNOWLEDGMENTS

Consultants/Contributing Writers

Science, Technology, and Society

Jon L. Harkness
Science Coordinator and Teacher
Wausau School District
Wausau, Wisconsin

Pam Helfers Riss
Science Teacher
Northlawn School
Streator, Illinois

Dorothy A. Tonnis
Science Department Coordinator
Logan Senior High School
La Crosse, Wisconsin

Ron Truex
S-T-S Instructor
Ocean Township High School
Oakhurst, New Jersey

Social Studies

David M. Helgren, Ph.D.
Department of Geography and
  Environmental Studies
San Jose State University
San Jose, California

Albert J. Vetrini, M.Ed.
Social Studies Teacher
Ocean Township High School
Oakhurst, New Jersey

Reading

Joan Develin Coley, Ph.D.
Director of Graduate Reading Program
Chair, Education Department
Western Maryland College
Westminster, Maryland

Patricia R. Love, M.Ed.
Adjunct Teacher
Western Maryland College
Westminster, Maryland

Reviewers

Science

Dr. Alvin Cuff
School Coordinator
Stoddart Fleisher Middle School
South East Region District
Philadelphia, Pennsylvania

Judith A. Mayer
1990-1991 Resource Agent
New York Science, Technology, and
  Society Education Project
Department Chair
John Burroughs Junior High School
Yonkers, New York

Mary Nalbandian
Director of Science
Chicago Public Schools
Chicago, Illinois

Scott Stokes
Science Teacher
Laney High School
Wilmington, North Carolina

Social Studies

Peggy Altoff
Specialist in Social Studies
Division of Instruction
Maryland Department of Education
Baltimore, Maryland

John Bellone
Social Studies Teacher
Wrenn Junior High School
Edgewood Independent School District
San Antonio, Texas

Dr. Norman McRae
Former Director of Fine Arts and
  Social Studies
Catherine Ferguson Academy
Detroit Public Schools
Detroit, Michigan

Harvey Prokop
Former Social Studies Specialist
San Diego Unified School District
San Diego, California

CONTENTS

S-T-S PROBLEM SOLVING
Dealing with the Issues

A Case Study: The Great Cereal Survey	4
Understanding S-T-S	7
"Campaign Will Seek Child Nutrition Labels" *The New York Times*	11
Getting Involved in S-T-S Issues	13
• Analyzing the Issue	13
• Gathering Information	17
• Making a Decision	20
• Taking Action	21
An Invitation	23

EXTINCTION of Living Things

Overview: Extinction of Living Things	24
Discovery	34
1 **"Foxes Hunted to Save Rare Bay Birds"** *The San Francisco Chronicle*	36
2 **"No Room to Roam"** *The Los Angeles Times*	40
3 **"Should We Downlist Our National Symbol"** *American Forests*	47
4 **"Agency Sets Owl Acreage at 6.9 Million"** *The Oregonian*	52
5 **"Alternative Energy vs. The Rain Forest"** *The Boston Globe*	56
6 **"Barnyard Rarities Get Their Day in Sun at Beltsville Show"** *The Baltimore Morning Sun*	61
7 **"Pessimism Is Growing on Saving Pandas From Extinction"** *The New York Times*	64
8 **"Horns of a Dilemma"** *Citizen Register—Westchester County, New York*	70
9 **"Closing in on Wild Bird Trade"** *The Washington Post*	73
Wrap-Up	78

HUMAN Populations

Overview: Too Many People, Too Little Space	**80**
Discovery	**90**
1 "Groups Unite to Point Out Hazards of Overpopulation" *The Los Angeles Times*	92
2 "Running Out of Room" *The Toronto Star*	95
3 "Busting the Boom: Population Control Works" *WorldPaper*	100
4 "Norplant Renews Debate Over Forced Contraception Reproductive Rights" *The Morning Call*	105
"Birth Control or Woman Control?" *Charlotte Observer*	109
5 "Abortion Issue Divides Advocates for Disabled" *The New York Times*	111
6 "U.S. Laws Under Attack" *The San Francisco Chronicle*	115
7 "An Investment in American Citizenship" *The Washington Post*	119
8 "Administration on Aging Announces The National Eldercare Campaign" *Aging*	123
Wrap-Up	**126**

HUMAN HEALTH and Disease

Overview: Healthy Living — 128

Discovery — 138

   1 "U.S. Reorganizes Nutrition Advice" — 140
      The New York Times

   2 "Schools Relax TB Deadline, Let Kids In" — 145
      Newsday—New York, New York

   3 "Doctors' Group Urges Tough Laws on Smoking" — 148
      The Los Angeles Times

   4 "Senate Panel Hears of Pesticide Harm Abroad" — 152
      The Los Angeles Times

   5 "Better Safe Than Sorry?" — 156
      Time

   6 "HIV Tests in the Health Profession" — 160
      The Washington Post

   7 "Behind Animal Testing Is Profit Motive" — 164
      Roanoke Times & World News

     "Think Animal Testing Is Inhumane? Explain It to the Sick, Dying" — 166
      Charlotte Observer

   8 "Laying Siege to a Deadly Gene" — 169
      Time

   9 "Group, Legislations Target Women's Health Research" — 173
      The Detroit News

Wrap-Up — 176

WORLD Food Resources

Overview: World Food Resources	178
Discovery	188
1 **"Famine in Africa, the Other Desert Crisis"** The Arizona Republic	190

**"World Hunger Is Persistent But Not Inevitable,
        Says New Report"**
        The Christian Science Monitor
	192

"In a Changing World, Little Has Changed for the Hungry" The St. Petersburg Times	195
2 **"Hunger Said to Afflict 1 in 8 American Children"** The Washington Post	197
"Survey: 160,000 Georgia Children Go Hungry" Atlanta Journal	198

3 **"Sustainable Agriculture More Pragmatic
        Than Organic Farming"**
        St. Paul Pioneer Press
	203

**"Organic Farming Is the Solution to Residue
        Pesticides, Expert Says"**
        St. Paul Pioneer Press
	205

4 **"State's Growth Threatens Way of Life in Rice Towns"** The Los Angeles Times	207
5 **"Irradiated Food Coming, But Not Without Protest"** The New York Times	213
6 **"Genetic Diversity Prevents Blight, Spread of Famine"** The Star Ledger—Newark, New Jersey	220

7 **"Scientists Use Gene Alterations to Make Crops
        Resistant to Infestation"**
        The Los Angeles Times
	224

Wrap-Up	228

SKILLS FOR GATHERING INFORMATION

Brainstorming	**231**
Making a Scrapbook	**232**
Keeping a Journal	**233**
Writing a Letter to Request Information	**234**
Using the Telephone Effectively	**235**
Organizing Data in a Data Table	**236**
Using the Catalog in the Library	**238**

SKILLS FOR TAKING ACTION

Writing a News Release	**239**
Making a Persuasive Speech	**240**
Writing a Letter to a Member of Congress	**242**
Setting Up an Awareness Fair	**243**

COMMUNITY RESOURCE DIRECTORY

General Organizations	**244**
Organizations for Young Adults	**252**
Publications of Interest	**253**
Government Resources	**256**
Science Museums and Zoos	**262**
Career and Trade Organizations	**266**

ACKNOWLEDGMENTS	**268**
ART AND PHOTO CREDITS	**270**
GLOSSARY	**271**
INDEX	**273**

"Opinions cannot survive if one has no chance to fight for them."
—Thomas Mann, 1924

S-T-S PROBLEM SOLVING
Dealing with the Issues

IN THIS MODULE

A Case Study: The Great Cereal Survey	4
Understanding S-T-S	7
"Campaign Will Seek Child Nutrition Labels" from *The New York Times*	11
Getting Involved in S-T-S Issues	13
• Analyzing the Issue	13
• Gathering Information	17
• Making a Decision	20
• Taking Action	21
An Invitation	23

Some ninth graders in Wisconsin want to find out which breakfast cereal is the best choice. Children everywhere need to understand what is in foods of all kinds so that they can make better food choices. Both problems have something to do with science and technology. People disagree about what the solutions should be. How can they decide what to do?

People find it hard to solve problems related to science and technology. They often don't know where to start or how to proceed. This module will give you a method for dealing with such problems. You will also have a chance, in the modules that follow, to use this method to understand the problems described in news stories from around the country. Then, you'll have a chance to use the method to tackle science and technology problems right where you live and go to school. Along the way you'll want to understand more about nature and what it takes to be an effective citizen.

A CASE STUDY: THE GREAT CEREAL SURVEY

Choosing a breakfast cereal is confusing. Supermarkets have hundreds of brands. Advertising leads us to believe that each brand is best. We know that a good breakfast is a healthful one. How can we be sure about our cereal choice?

Science Lesson Identifies a Problem

Recently, ninth grade students at North Crawford High School in Gays Mills, Wisconsin, were studying the

The students in Mr. Jurgensen's science class worked with a computer literacy class to design a computer spreadsheet program to analyze their data.

topic of food with their teacher, Wayne Jurgensen. The students learned that breakfast cereal is an important kind of food. It provides good nutrition at a low cost—usually.

Are some cereals more healthful than others? How do the prices compare? Mr. Jurgensen asked the students to look for answers to these questions. They did, and they were surprised at what they found.

Mr. Jurgensen gave the class five different cereal boxes to examine. He pointed out to the students the nutrition labels. Nutrition labels, by law, are printed on all cereal boxes. He also told the students the price of each of the brands of cereal.

Mr. Jurgensen explained that cereal is one of the least expensive sources of the protein that people need in their daily diet. Meat has protein, too, he explained, but the protein in cereal costs less than the protein in meat. The class decided to find out which cereal gave the "best deal" on protein.

Students used the label information and the price for each brand to calculate the cost per gram of protein for each of the cereals. They came up with some unexpected answers. For some cereals, the cost per gram of protein was only 2 or 3 cents. For other brands, the cost per gram of protein was more than 25 cents—quite a difference!

Students Want More Information

The students were surprised by the results and wanted to know more. They wanted to check their favorite brands of cereal as well as other brands they had seen on television commercials. Also, the students wanted to analyze other information shown on the nutrition labels. Mr. Jurgensen asked the students to bring in a few more empty cereal boxes.

As their collection of cereal boxes grew, students became more and more interested in the project. They decided to study the total nutritional content of cereal, as well as nutritional "cost". The class added the analysis of calories, carbohydrates, sodium, and sugar to their cereal survey.

Word about what the class was doing got around the school. Other students donated boxes, and other teachers stopped in to express interest in the project. Soon the class had 109 cereal boxes, all different, and they had a major project on their hands.

Students Gather Data

When the students gathered the data from all the boxes, they had over 1,500 pieces of information to work on. Too much, they thought. One student said, "It would have taken us till summer just to type up all that stuff on a typewriter." And it would have taken that long if they had not found a way to handle all of the information.

Technology as a Tool

At the same time that the students in the science class were working on what became the "Great Cereal Survey," another group of students happened to be studying spreadsheets in Joan Davig's computer literacy class. A **spreadsheet** is a computer program that organizes numerical data for computing desired calculations and making overall adjustments based on new data. Mr. Jurgensen suggested that the classes join forces to design and use a spreadsheet to analyze the cereal data. With help from Mrs. Davig's computer literacy class, the students in the science class entered their cereal data and gave the instructions to the computer for the calculations they needed.

The students had the computer print a list that showed for each of the 109 brands of cereal the following kinds of information:

- cost per serving
- cost per gram of protein
- grams of sugar per serving
- grams of carbohydrate per serving
- grams of protein per serving
- milligrams of sodium per serving
- calories per serving

Surprising Results

Now the students could really compare cereals. Here are some examples of what they found:

- The cost per gram of protein is over 30 times higher for some cereals than others.
- Some "low calorie" cereals do not contain much protein—they have little nutritional value of any kind.
- Some cereals contain too much sugar or too much salt for good health, according to the recommendations of many nutritionists.
- Brands that seem best from television commercials often are not very good cereals because of their nutritional content or cost.

Results Shared with Others

As students and staff throughout North Crawford High School saw copies of the "Great Cereal Survey" report, word spread into the community. The local newspaper interviewed class members and published an article about the project. In the article, the students offered a copy of their report to residents of the community and several people responded.

The students in Mr. Jurgensen's class, and others who read their report, are no longer confused about what cereal to buy. They know how to select the best cereal for the lowest price. They will also be able to apply what they learned about nutrition labels to selecting foods throughout their lives.

UNDERSTANDING S-T-S

What Is Science?

Science is a human activity directed toward trying to understand nature. The goal of science is to understand how every bit of matter and energy, from the tiniest parts of atoms to the entire universe, works. People go about "doing" science by *asking a question*—a question about how something in nature works. Next, an answer is sought by *observing* nature. That is, science answers include only what can be observed by seeing, hearing, smelling, or the use of the other senses. Often, instruments are used in science to assist the senses. For example, radar extends the human ability to "see" distant objects.

Based on their observations, scientists *seek patterns* in nature. They use these patterns to *create models* that explain how parts of nature may work. The models of science vary in their form. Some are mathematical equations, such as the law of gravity. Others are actual physical models, such as an atom represented by a few colored balls and wires.

SOME PROCESSES OF SCIENCE

- Asking questions
- Observing
- Seeking patterns
- Creating models

This description of science should sound familiar to you. In your science classes you may have discussed the "scientific method." The scientific method is a set of processes that one uses when "doing" science. The scientific method is another way of defining what science is and defining what activities are scientific.

THE METHODS OF SCIENCE

- Questioning
- Gathering information
- Forming a hypothesis
- Experimenting
- Collecting data
- Analyzing data
- Concluding

Who Can "Do" Science?

Who is, or can be, a scientist? Anyone is or can be a scientist at any time, if he or she:

1. has a question about how something in nature works, and
2. seeks an answer by observation of things and events in nature.

Science is not limited to people who have college degrees in science or people who work in laboratories. At some time or another, everyone acts in a way that can be called scientific. Think of the students in Mr. Jurgensen's class. What did they do that was science?

What Is Technology?

Technology is a human activity. **Technology** involves people applying knowledge, tools, and skills to make and do things that are useful in life. Technology is separate from, but related to, science. The role of science is to understand things; the role of technology is to apply things to our lives. The goals of technology are:

1. to solve practical problems and
2. to extend what humans can do.

Scientific experiments can be carried out in a variety of settings.

Who Can "Do" Technology?

Who is, or can be, a technologist? Just as anyone can do science, so can anyone do technology. At some time, everyone "invents" a way to solve a problem or a device that is useful. Name or describe something that you or a friend has invented.

Some people make technology their careers. **Engineers** are highly trained in science, mathematics, and technology. They plan, design, and help build complex devices such as highway systems and assembly lines for manufacturing. People who "tinker around" and invent something are also technologists. Perhaps the best example of a "tinkerer" is Thomas Edison. Edison seemed to rely on trial and error in his work. While his methods were not very fancy, his inventions, such as the light bulb, changed the world.

For example, the students in Wisconsin used technology in the form of a computer. The computer spreadsheet program extended their ability to do calculations. It allowed the students to do thousands of arithmetic problems with great speed and accuracy. Without computer technology as a tool, it would have been impossible to compare the cereals in the available time.

Science and technology start with different questions. Science starts with a question about the natural world. Technology starts with a question about what needs to be done, or what could be made to improve a situation.

How Do Science and Technology Interact?

Most people think that science comes first, and then technology takes over and turns the understanding into something useful. That kind of interaction happens often, but not always. Sometimes, technology comes before science. For example, ancient healers, or shamans, crushed plants to make brews that they fed to sick people. Through trial and error, they found some very useful medicines. No scientific understanding of the human body, or of the plants, was involved with that technology. Yet, we still use many of those medicines. For example, belladonna, from nightshade plants, was used traditionally to relieve stomach cramps. Today, belladonna is used to make certain prescription drugs.

Think of science and technology as two human activities that feed each other. On one hand, new understandings of science can be used to create a new technology. On the other hand, the tools developed by technology are used by scientists to gain further understanding of nature.

What Is Society?

Since the beginning of recorded history, people have lived together in groups called communities. In early times, communities were isolated from one another. People were involved only with other members of their own local community. As the world's human population grew, communities came in contact with one another and began to form nations. Then people became involved with other people beyond their local area and community.

Today, over five billion people live on earth. Each person is involved in several levels of organization. In the United States the levels include town or city, county or parish, state, region, and nation. All of the communities of people at all those levels are called **society**.

How Do Science, Technology, and Society Interact?

Since science and technology are carried on by people who are members of society, there are interactions among science, technology, and society. You have learned that science and technology affect each other. Science and technology also affect society and society affects each of them.

SCIENCE OR TECHNOLOGY?

Can you distinguish science from technology? Check yourself by classifying each of the following questions as answerable by either science or technology.

1. What are the functions of protein in human nutrition and health?
2. How can the amount of protein in different brands of cereal be measured?
3. How do seed plants reproduce?
4. How can farmers increase the yield of their food crops?
5. What are the characteristics of the virus that causes AIDS?
6. How can the spread of AIDS be stopped?

Think about the problem with breakfast cereals at North Crawford High School. The members of Mr. Jurgensen's class were acting as scientists. The purpose of their work was to understand more about the properties of breakfast cereals.

Technology had made available the ability to analyze and compare several properties of over 100 brands of cereal. Therefore, the students were able to use technology to "do" science.

At the same time, the students were members of the society of their school, city, state, and nation. The information that the students discovered through the use of science and technology was used by many other people.

What Is Scientific Literacy?

To say that a person is *literate* means that he or she can read and write. Literacy is the ability to read and write. Citizens of a democracy participate in their government by getting information about issues, exchanging opinions, debating, and voting. It is difficult to get involved in these things if you are not literate.

Our society has a need for another kind of literacy, scientific literacy. **Scientific literacy** is the ability to use the understanding of science and technology to solve everyday problems and make decisions about issues in our society. The need for scientific literacy is based on the fact that more and more of the decisions that citizens make as they govern themselves involve science and technology. For example, people now realize that technologies developed in the past can haunt us now and in the future. Water, air, and land pollution is the result of technologies such as factories, automobiles, chemical fertilizers, and pesticides. Today, our society is

Reading newspapers is an easy way to gather information about the many issues facing society today.

faced with making decisions about how to clean up pollution in order to preserve our planet.

What's in the News?

People become aware of S-T-S issues in a variety of ways. Often, the first awareness comes through the news media: newspapers, magazines, radio, and television. This book contains a collection of newspaper and magazine articles from around the United States. Each article discusses an issue related to science and/or technology and society. As you read them, you will begin to comprehend what kinds of knowledge will help you understand each issue. That knowledge is part of scientific literacy. The issues raised in the articles will get you involved in learning how to solve S-T-S problems—and by becoming involved, you will develop your scientific literacy.

> Here is an article from a newspaper in New York. The story takes place in Washington, D.C., at a conference sponsored by a group called Public Voice for Food and Health Policy. As you read it, look for the interactions among science, technology, and society. Also, think about how the citizens at the conference could use their scientific literacy.

Campaign Will Seek Child Nutrition Labels

Private group takes cue from F.D.A. chief's idea

By Warren E. Leary
FROM: THE NEW YORK TIMES
JANUARY 14, 1992

Seizing an idea advanced by the Commissioner of Food and Drugs, a private group said today that it was starting a voluntary campaign to develop food nutrition labels designed to help children better understand what is in their food.

The group, Kidsnet, outlined its plans at a conference attended by the commissioner, Dr. David A. Kessler.

The nonprofit information group, which is supported by charitable foundations and the cable and broadcast industries, said it would convene conferences of experts and develop food labeling aimed at children from 6 to 12 years within a few months. The group said it would then try to persuade food companies, whose reaction to the proposal has been cautious, to add the children's information to existing labeling on products aimed at young consumers.

Food industry spokesmen said, however, that while the aims of the Children's Food Labeling Initiative [a group proposing new legislation] were praiseworthy, separate labels might be impractical, unnecessary, and confusing.

Making Children Aware

Dr. Kessler, a pediatrician and parent, has advocated [supported] a children's labeling and nutrition education initiative privately, partly because his agency has neither the federal authority nor the money to require and police such a program.

"We were inspired by Dr. Kessler's notion of a unique food label for kids," Kidsnet's executive director, Karen

Jaffe, said at the conference sponsored by Public Voice for Food and Health Policy, a consumer advocacy group [a group that speaks or writes in support of a position or cause]. The idea of helping children to become aware of what is in their food and to be more healthful consumers "struck a chord with us," she said. . . .

"We clearly have a few ideas about the label," Ms. Jaffe told the conference. "We think it should be bold and simple, with the emphasis on graphics rather than words. We also have notions of what it should not be—an endorsement [approval] of any kind, a replacement for the regular label, judgmental about food, or a way to make claims not otherwise allowed."

The Grocery Manufacturers of America, a trade association representing 130 companies that make 85 percent of the food and grocery products sold in the country, said this was not a good time for another labeling proposal.

> ❝*Food industry officials are hesitant about twin labels.*❞

Jeffrey Nodelman, a vice president of the trade group, said that because of the Nutrition Labeling and Educational Act of 1990, the industry was in the process of updating all the nation's food labels for the first time in 30 years. After a year of work the industry and the Food and Drug Administration have yet to agree on even the format of the new labeling requirement, he said, and more new proposals would only complicate the process.

"The new law doubles the diet and health information required on labels, and we're still trying to figure out how to get all of that into a limited space," Mr. Nodelman said. "We don't believe a voluntary second label, in addition to the new one, is the right answer.

Public Voice, a consumer group that sponsored the conference where Kidsnet outlined its plans, also had reservations about the effectiveness of a voluntary program. Ellen Haas, the group's executive director, said a children's nutrition education program would need the financial and regulatory backing of the F.D.A., the Agriculture Department, the Health and Human Services Department and other agencies to be effective.

High-Fat School Lunches

"Nutrition education is not just a public relations effort and must be made a priority in President Bush's 1993 budget in order to fully succeed," Ms. Haas said. "Public Voice has documented the dangerously high fat content of meals in school lunches and the glaring absence of nutrition information provided to kids," she added.

The National Food Processors Association said it was too early to endorse the concept of special food labeling for children. Noting that the food and drug agency was already in the process of changing all food labels, the association president, said: "The outcome of the F.D.A.'s ongoing research into different food label formats could mean that no special child's labeling is needed, since the whole idea is to come up with a format that is simple to understand and easy to use."

GETTING INVOLVED IN S-T-S ISSUES

You have just read a newspaper article called "Campaign Will Seek Child Nutrition Labels." The situation described in the article is an example of a Science, Technology, and Society, or S-T-S, issue. It fits the definition of an S-T-S issue because:

1. There is a problem related to science and/or technology.
2. People strongly disagree about the solution to the problem.

Often, S-T-S problems are difficult to deal with. Citizens don't know where to start or how to proceed. However, S-T-S problems can be understood. You can become better informed about an S-T-S problem. You can form your own opinion about what should be done. You can do something about the problem as a student, a citizen, or a member of a family. Anyone can learn how to go about it. All it takes is learning a method for attacking S-T-S problems and practicing using the method. Along the way you will find out about some of the many ways that citizens participate in our government.

This section of the module will give you the opportunity to learn a method for dealing with S-T-S issues. You will practice the method for the case of the cereal survey at North Crawford High School and the case described in the article "Campaign Will Seek Child Nutrition Labels." The problem solving method has four steps:

1. Analyze the Issue
2. Gather Information
3. Make a Decision
4. Take Action

The situation described in "Campaign Will Seek Child Nutrition Labels" may affect everyone. The students at North Crawford High School had a big job to analyze labels for only one kind of food, cereal. Adults, as well as students, need to make use of nutrition labels, and the task of analyzing so many cereal labels would not be any easier for adults than it was for the students. It is important for everyone to understand what is in their food. Later, you will be invited to apply the method learned here to other S-T-S issues.

ANALYZING THE ISSUE

S-T-S PROBLEM SOLVING

1. Analyze the Issue   **2.** Gather Information   **3.** Make a Decision   **4.** Take Action

The article "Campaign Will Seek Child Nutrition Labels" suggests a question: How well do you understand what is in your food? Therefore, the problem of nutrition labels affects you.

The first step in dealing with any S-T-S problem is to analyze the issue. To analyze means that you break the problem down into small understandable pieces. After looking at each piece, you will put them all together again so that you can explain the problem to others. The method of analyzing an issue has five steps:

1. Identify the problem.
2. State the issue.

3. Identify the players.
4. Describe the players' positions and attitudes.
5. Describe the proposed solutions.

Identify the Problem

To get started, take a look at the source of your information. On a separate sheet of paper, write down the name of the source of information. In the example, the source is the newspaper from which the article "Campaign Will Seek Child Nutrition Labels" was taken. Include the name of the writer.

Now you are ready to identify the problem. A problem is a situation that is perplexing or troubling or difficult. Read the article again and write down the answers to these questions:

- What is the problem?
- How did the problem come to exist?
- How are science, technology, and society each involved with it?

Perhaps you noticed while answering the questions that it is difficult to "stick to the facts" and not include your own opinion. Putting aside your own opinions while examining a problem is called *being objective*. In this case, this means looking at the problem of children's nutrition labeling from the viewpoints of the manufacturers as well as the government agency and the consumer advocacy group. Being objective takes practice, and this module will give you many opportunities for that.

State the Issue

An **issue** is a question to be resolved. It is a restatement of the problem in the form of a question. Remember that each person quoted in "Campaign Will Seek Child Nutrition Labels" had a different idea about how to handle food nutrition labels. To state the issue:

- Review the problem that you wrote down.
- Rewrite the problem in the form of a question.

Identify the Players

The **players** in any S-T-S issue are the people who are involved. The players may include individuals, groups of people, or organizations with members and employees. The players each have a role related to the issue. The role of each player is how the player behaves in relation to the issue. The relationship may be determined by the player's job, his or her responsibility, gender, ethnic group, or some other personal interest. For example, family members, who shop for groceries and are concerned about good nutrition for their families, may be active players in the case of deciding on nutrient labels for food.

To identify the players, return to the article "Campaign Will Seek Child Nutrition Labels" to find the information needed to complete these steps:

- Write down the S-T-S issue.
- Write down the name of each player mentioned in the article.
- Write down the role of each player.

You may wish to organize your thoughts in a chart like the one below. Copy it onto a separate sheet of paper.

It is not easy the first time around to include everyone who may be involved in an S-T-S issue. Compare your chart with those of other students to make sure that you identified all the players involved in the "Campaign Will Seek Child Nutrition Labels" article. You may then want to refine your list. For example, some people might not include as players the children who choose some of their own foods. Those children are involved, because the nutrient content in their food affects their health.

Describe the Position of Each Player

The **position** of a player is how he or she feels about the issue. The position is revealed by what a player does or says about the issue. Two things contribute to a player's position on the issue: the player's beliefs and the player's values. A **belief** is an idea that the player holds to be true, whether it is really true or not. A **value** is a quality or principle that is considered worthwhile.

What a person believes and values is often determined by that person's role in society. A person who is a member of one ethnic group may have certain experiences that members of other groups have not had. These experiences affect beliefs and values. Members of one religious group may also have different beliefs and values than members of another religious group. Gender too, that is whether a person is male or female, may affect what position a person takes on a particular issue.

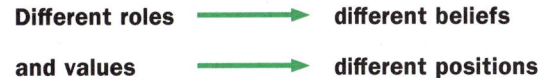

Now, look back at the article "Campaign Will Seek Child Nutrition Labels" and write on a separate sheet of paper the following:

- State the issue in the form of a question.
- State the names of the players.
- Describe the position of each player.
- State one belief and one value of each player that helps explain his or her position.

You may wish to draw a chart, like the one below, to organize your ideas.

Issue:			
Player	Position	Belief	Value

It was probably harder for you to describe beliefs and values than to identify the players, because beliefs and values are not stated in the article. You have to infer, or conclude, what the beliefs and values are from the evidence presented in the article. Your statements are called inferences. An inference is a conclusion made from observations. Chances are that you and your classmates have made different inferences about the players in "Campaign Will Seek Child Nutrition Labels." You may have said that John R. Cady, president of the National Food Processors Association, believes that no special child's labeling is needed. A classmate may have said that Cady really does believe that special labels for children are needed, but that he is afraid to say so. Try working out some of these differences in discussions with other students in your class.

Describe Solutions

The final step in analyzing the issue is to describe the solutions proposed by the players. A solution is what a player in an S-T-S issue thinks should be done to solve the problem. Some players may not offer a solution. The position and solution of other players may be one and the same.

You will need a separate sheet of paper. Also gather all of the information you have written down about the players' roles, positions, beliefs, and values.

Review the article "Campaign Will Seek Child Nutrition Labels" and then follow these steps:

- Write down the issue.
- Record the names of the players.
- Write down the solution or solutions proposed by each player.

You may wish to use a chart like the one below to record your ideas.

Issue:	
Player	Solution

Now you have completed the first step in S-T-S problem solving, analyzing the issue. Perhaps you feel that your results are not complete in some ways. The problem and issue may still be unclear to you. You may suspect that more players are involved than you recorded. You may have been at a loss to identify beliefs and values for some of the players. The list of solutions may seem too limited to solve the problem. The reason is that you have used only one source for your information. The source of information from which people get their first understanding of an S-T-S issue is usually incomplete.

In analyzing "Campaign Will Seek Child Nutrition Labels" you probably have thought about additional information that would be helpful. That need leads to the next step in S-T-S problem solving: gathering information.

GATHERING INFORMATION

S-T-S PROBLEM SOLVING
1. Analyze the Issue   **2.** Gather Information   **3.** Make a Decision   **4.** Take Action

This step in S-T-S problem solving will get you beyond the information provided in the original source. Gathering additional information can be time-consuming. So it is important to identify, as exactly as possible, what information is needed.

Four things need to be done when gathering information:

1. Identify the *kinds* of information needed.
2. Identify possible *sources* of that information.
3. Plan a *method* for getting the information.
4. *Evaluate* the information once it is gathered.

Identify Kinds of Information

As always, the best way to get started is to return to your statement of the issue. On a separate sheet of paper, make a list of kinds of information that will help to answer the issue question. The following hints will guide you in making your list.

- Gather information that applies to the issue.
- Gather information about similar situations.
- Gather information that will help you make a decision about the issue.
- Gather information about unheard players and their roles, positions, and solutions.

When gathering information about the unheard players, make a list of people who might be involved in the issue but were not mentioned in the source. For example, a nutrition problem with school lunches was mentioned in the newspaper article "Campaign Will Seek Child Nutrition Labels." Since many students have meals at school, that may be part of the issue. Think about the people connected with school lunches who might be sources of additional information. With any problem, an understanding of all of the groups that make up society will be useful when identifying the group of players that may have been overlooked in the source material.

You can use the issue presented in the article "Campaign Will Seek Child Nutrition Labels" to practice information gathering. Use the hints above to guide you in writing down at least three questions about the child nutrition label issue that you would like answered. One question might be, "Have children ever been tested about how well they understand what is in their food?"

Identify Sources of Information

Now that you know what information is needed, the next step is to figure out where to get the information. Start by thinking about how science, technology, and members of society can provide information. For example, science can provide information about the effects of certain pollutants on human health. Technology, in the form of tools and instruments, can be used to provide the measurements that will sharpen your observations. People, as members of

Interviewing a person is a good way to gather information.

governments, agencies, and special interest groups, can provide expert information on the issue.

When approaching people for information, make sure that you distinguish between those who are players in the issue and those who are not. People directly involved in a controversy may find it hard to be objective. People not directly involved may be a better source of information.

Society also can provide information in the form of *records*. Whenever people organize into groups for a purpose, such as governing, conducting business, or special interests, records of activities are made and saved. The records can be in the form of printed words, audiotapes, or videotapes. For example, it might be useful to have a copy of the new federal law, the Nutrition and Labeling Educational Act of 1990, mentioned in the article "Campaign Will Seek Child Nutrition Labels."

Now practice identifying sources of information that would be helpful in

> **B**eginning on page 231 of this book, you will find instructions on how to carry out some of these Skills for Gathering Information. Use them when you are ready to gather your own information.

understanding the issue described in "Campaign Will Seek Child Nutrition Labels." For each bit of information that you want, write down on a sheet of paper the source from which you might be able to get it.

Methods for Gathering Information

After you have decided on your sources, you can decide *how* to get the information from those sources. Gathering information from an individual person may require an interview. Interviews can take place in person or by telephone. To get information from a group of people, a public opinion survey could be used.

Gathering information from records is complicated because records exist in a variety of forms and in a variety of locations. Often the best place to start is the library. If you want to find magazine articles about your issue, try the *Readers' Guide to Periodical Literature*. The *Readers' Guide to Periodical Literature* is a set of reference books in which you can look up topics of interest. Many libraries also have computer databases in which you can search for newspaper articles and other information. Reference librarians are experts in gathering information, and most are quite willing to help people in their searches.

You may also want to gather scientific information directly. For example, you could find out from a nutritionist, a food scientist, what should be in a person's daily diet. Then you could test how easy or difficult it is, using nutrition labels and other information, to find out how healthy—and how expensive—some typical daily diets are. Your class's meals for a day could be used in a sample study.

Now practice on the issue described in the article "Campaign Will Seek Child Nutrition Labels." Look back at your list of the information you said you would want and the sources from which you would get it. Write down on a sheet of paper the methods you would use to get the information you identified.

Evaluate the Information

Information should be evaluated as it is being collected as well as after the process is completed. To be useful, each bit of information collected should involve a "yes" answer to one or more of these questions:

- Does the information really apply to the issue?
- Does the information concern a situation truly similar to the one you are investigating?
- Will the information help you make a decision about the issue?
- Does the information add to the list of players involved with the issue?

MAKING A DECISION

S-T-S PROBLEM SOLVING

1. Analyze the Issue   **2.** Gather Information   **3.** Make a Decision   **4.** Take Action

The third step in the S-T-S problem solving method is to make a decision. When you make a decision, you conclude what the best solution of the issue is. It is best not to jump to conclusions or make a decision about an issue until the issue has been analyzed and additional information has been gathered.

When making a decision, four questions should be answered:
1. Do you have the information needed to make a decision?
2. What are the options?
3. What are the consequences of each of the options?
4. Which option is best?

Do You Have the Information to Make a Decision?

To answer this question, you have to look back at the information that was gathered. Often a cause and effect relationship is at the heart of the problem. For example, in the article "Campaign Will Seek Child Nutrition Labels," some players claimed the effect was that children do not understand well enough what is in their food. They pointed to the lack of nutrition education and information for children as the cause.

For any issue, identify the cause and effect relationship and any evidence, or facts, that back up your statement. Also, state the source of the information that backs up your cause and effect statement. You may want to copy the chart below. Then try filling in the rest of the chart.

What Are the Options?

Options are different ways that the issue can be resolved. Look at the issue you wrote down. It should be a question. The options are possible answers to that question. The options suggest what can be done about the problem.

The Kidsnet group decided to seek child nutrition labels as its option for resolving an issue. Imagine that you were a member of this group. List on a sheet of paper some other options that Kidsnet should consider to help children better understand what is in their food.

Issue: Should special nutrition labels for children be developed?	
Cause and effect statement: The lack of nutrition education and information causes children not to understand well enough what is in their food.	
SOURCE OF EVIDENCE	SUPPORTING EVIDENCE

What Are the Consequences of Each Option?

Consequences are the results or effects of an action. Different options usually have different effects. Two options may both resolve the issue, but they can have different side effects. A **side effect** is an effect in addition to the one that is desired. Sometimes the cost of one option is much greater than the cost of the other options.

Each of the consequences may be classified as a benefit or a cost. A **benefit** is a positive, or good, result; a **cost** is a negative, or bad, result.

Practice identifying benefits and costs using the list of options that you made. On a sheet of paper, create a chart like the one below to record one benefit and one cost for each option.

OPTION A:	
Benefit	Cost
OPTION B:	
Benefit	Cost

Which Option Is Best?

The final step in decision making is to pick the one option that seems best to you. This requires you to make a judgment. Judging the options involves analyzing trade-offs. **Trade-offs** are compromises that are arrived at by weighing the benefits against the costs. One way to weigh the trade-offs is to assign number scores to each benefit and cost. Here is an example of one way to do it:

EFFECTS	POINTS
A large benefit	+10
A slight benefit	+ 5
A slight cost	– 5
A large cost	–10

You can figure out a total score for each option by adding up the points. Next, use the total scores to rank the options. The number one position in the ranking goes to the option with the highest score. The option at the top of your list is your solution to the issue.

TAKING ACTION

S-T-S PROBLEM SOLVING
1. Analyze the Issue   **2.** Gather Information   **3.** Make a Decision   **4.** Take Action

Now that you have a solution in mind, you need to think about what you will do about it. An **action plan** identifies the things that you can do as an individual or as a member of a group in order to get your solution accepted by others.

Persuasion

Often, the major focus of an action plan is persuasion. **Persuasion** involves trying to convince people of a need for change. To get people to accept your solution, you have to make them aware of the problem, understand the need for action, and find value in your solution. You can do these things by sharing with people your analysis of the issue and the information that you gathered. You can get others to value your solution by sharing with them how you made your decision. Because all these previous steps will add to your ability to persuade, it is important that you keep accurate records of all your S-T-S problem solving steps.

Audiences

Who needs to be convinced of your position? All members of society who may be able to help reach the solution you have chosen. Your audience is the players you identified and all other members of society who might become players and take a stand that will support your goal.

Kinds of Social Action

There are many actions that you can take as a member of society. You can write a letter to a member of Congress, write a letter to the editor of your local newspaper, start a petition, picket, form a club for others to join, and so on.

In order to decide what action you want to take, the following six questions should be answered:

1. What actions can you take as an individual?
2. What actions can you take as a member of a group?
3. Which of these actions would be most effective in creating a change?
4. Which actions from question 3 would you be able to carry out effectively?
5. Pick the best action from list #4. Make sure to include a description of your audience.
6. Explain why this is the best choice for you.

Again imagine that you are a member of the Kidsnet organization. The organization has decided that child nutrition labels are needed on food packages. On a separate sheet of paper, answer the six questions as you think members of the Kidsnet group might. Answers to question 1 might include:

- Vote in the organization to promote child nutrition labels and forget about it.
- Send letters to the Grocery Manufacturers of America and the National Food Processors Association.
- Suggest that the organization invite discussion with the Food and Drug Administration.

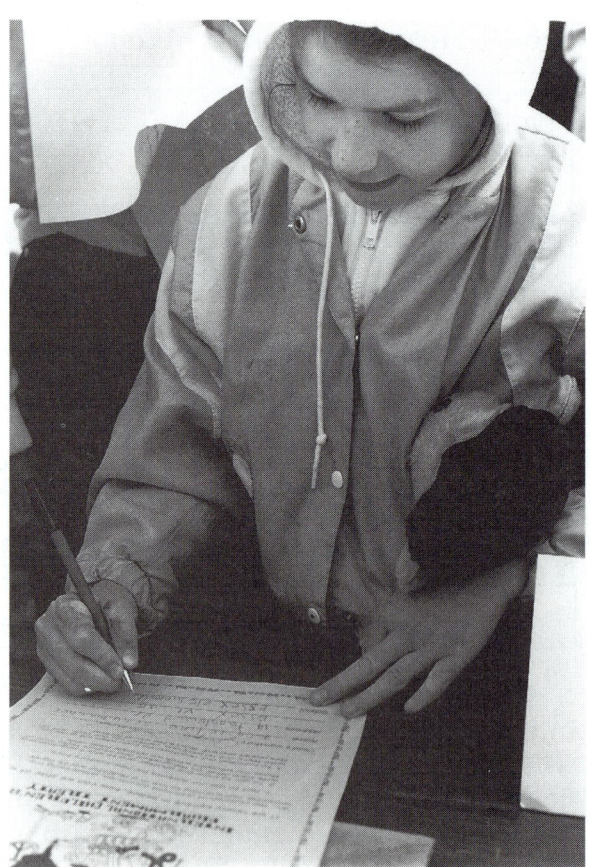

Circulating a petition is an effective type of social action.

Beginning on page 239 of this book, you will find instructions on how to carry out some of these Skills for Taking Action. Use them when you decide what action you will take.

An Invitation

Almost every day, new S-T-S problems are in the news. It is certain that the quality of life in the future will depend on our ability to solve these problems. To be an effective citizen, it is necessary to know how to deal with problems related to science and technology. As with any set of skills, you will get better at S-T-S problem solving with practice.

The story told in the article "Campaign Will Seek Child Nutrition Labels" and the story about the students who analyzed breakfast cereal labels and prices were used to introduce you to the skills. Now you are invited to sharpen your skills by applying this problem-solving method to newspaper and magazine articles that involve other issues.

The real test of your skills will come when you apply the method to an S-T-S problem or situation in your own life. That's what the students in Wisconsin did when they conducted the "Great Cereal Survey."

You may wonder how to start working on S-T-S problems where you live. Begin by thinking of some small problems. Small S-T-S problems can probably be found right in your school or neighborhood. Does your school cafeteria serve nutritious lunches? Will a new shopping mall being built destroy the habitat of wildlife? Are the schools crowded because the population is growing too fast? No doubt, you can think of many other problems.

Be scientifically literate. Use what you know about science and what you will learn by using this book to begin to find solutions to these problems. By doing this you will become a more effective citizen.

Students learn how Congress works by participating in a government workshop.

EXTINCTION
of Living Things

"... Nature is one and continuous everywhere."
—Henry David Thoreau, 1849

EXTINCTION of Living Things

IN THIS MODULE

Overview	27
Discovery	34
1 **"Foxes Hunted To Save Rare Bay Birds"** from *The San Francisco Chronicle*	36
2 **"No Room to Roam"** from *The Los Angeles Times*	40
3 **"Should We Downlist Our National Symbol?"** from *American Forests*	47
4 **"Agency Sets Owl Acreage at 6.9 Million"** from *The Oregonian*	52
5 **"Alternative Energy vs. The Rain Forest"** from *The Boston Globe*	56
6 **"Barnyard Rarities Get Their Day in Sun At Beltsville Show"** from *Baltimore Morning Sun*	61
7 **"Pessimism Is Growing on Saving Pandas From Extinction"** from *The New York Times*	64
8 **"Horns of a Dilemma"** from *Citizen Register (Westchester County, NY)*	70
9 **"Closing in on Wild Bird Trade"** from *The Washington Post*	73
Wrap-Up	78

EXTINCTION OF LIVING THINGS

Have you ever seen a tyrannosaurus? Perhaps you've seen models or fossils of this meat-eating dinosaur in a museum. Or maybe you have seen models of *Tyrannosaurus rex,* with its long, sharp teeth and powerful jaws, in a movie. But living tyrannosaurs have not existed on Earth for millions of years.

Dinosaurs were the dominant land animals for almost 160 million years. Then, over a period of several million years, they all died out, or became **extinct.** Many other kinds of organisms also became extinct along with the dinosaurs. Throughout the long history of life on Earth, there have been several times when large numbers of different kinds of organisms died out. In fact, more than 90% of all the species that have ever lived on Earth are now extinct.

Extinction is a natural process that occurs as conditions on Earth change. The climate may become hot and dry. Or huge sheets of ice may creep over the land, creating a frozen desert. An organism suited for one type of environment may not be able to survive the changes. New life forms with the characteristics needed to survive the new conditions, may evolve over time. Yet today, the process of extinction has many people worried. Normally, extinction occurs over extremely long periods of time—hundreds of thousands or even millions of years. However, human activities have changed the rate of extinction. According to some estimates, people are driving one species of either a plant or animal a day to extinction. Others believe many species are dying

Dinosaurs, such as the *tyrannosaurus rex*, ruled the earth for millions of years.

out every day. By the year 2000, people may be responsible for the extinction of as many as one million species of plants and animals.

Why Extinction Matters

If extinction is a natural process, why should we be concerned about it? After all, scientists estimate that there are between 5 and 30 million different plant and animal species on Earth. Just 1.4 million of these have actually been studied and described. Of these, people depend on only 1 percent directly for survival. Do the species we don't use or don't know about really matter?

All living things exist in a web of relationships with other living things. Organisms also interact with non-living parts of their environment. Plants depend on the sun's energy to grow. Animals eat plants or other animals. Fungi and bacteria break down dead plants and animals and return nutrients to the soil. These interactions all occur in a thin band of Earth's surface called the **biosphere.** The biosphere reaches from 8 to 10 kilometers above sea level to several hundred meters below sea level. On land, the biosphere extends as far down as microorganisms and plant roots are found—just a few meters.

To study the different interactions among living things and their environment, scientists divide the biosphere into units called **ecosystems.** Usually, ecosystems have some natural boundary that separates them from other ecosystems. So an ecosystem can be as large as a forest or as small as a drop of rainwater. No matter what their size, however, all ecosystems have certain characteristics in common. Within an ecosystem, there is a flow of energy and a cycling of materials from the non-living environment, through the bodies of living things, and back to the non-living environment.

Organisms have fairly specific roles and relationships in ecosystems. Grass in a meadow is a producer, capturing the sun's energy and using it to create **organic compounds.** A grasshopper in a meadow is a consumer. It cannot make its own food. Instead, the grasshopper feeds on the organic compounds and stored energy in the grass. A frog eats the grasshopper and in turn may be eaten by a snake. Both of these animals are also consumers. At the top of this food chain is the final consumer, a hawk. When the hawk or any of the other organisms in this food chain die, decomposers, such as bacteria, break down the body tissues and release the organic compounds into the soil. The cycle begins again as the organic compounds fertilize the grass.

Imagine what might happen to an ecosystem if one or more species in it became extinct. The loss of one species can set off a chain of events in which many or most of the relationships in the ecosystem are destroyed. One estimate is that for every one plant species that disappears, between 20 and 40 animal species that depended on it, either directly or indirectly, may also disappear.

Think about the following example. In Kuala Lumpur, an area in Malaysia, people dug limestone out of caves and drained swamps to use the land. By doing this, they destroyed the nesting and feeding sites of a species of bat. They did not know, however, that this particular bat is responsible for pollinating one of Southeast Asia's most important fruit tree species. Now this fruit crop is threatened because there are fewer bats to pollinate the trees.

An eagle is a final consumer in the food chain.

People often take the interactions among organisms for granted, but our survival depends on them. We need forests to take carbon dioxide from the air and replace it with the oxygen we breathe. We need certain insects to pollinate food crops. Other organisms, such as insects and invertebrates, maintain the soil fertility that allows plants to grow. Unfortunately, the importance of a species to an ecosystem may not be discovered until after the species is extinct or nearly extinct and people become aware that its absence has harmed the ecosystem.

Extinction and Natural Resources

Before 1960, a child with leukemia, a cancer of the blood, had a one-in-five chance of survival. Today the odds of survival are four-in-five. The reason: two important drugs developed from a tropical rain forest plant called the rosy periwinkle. Many important medicines have been developed from chemicals extracted from plants. What would have happened if the rosy periwinkle had been driven to extinction before its valuable properties were discovered?

Extinction results in the loss of valuable resources that could improve the quality of life for people around the world. One scientific study found that over 1,000 plants used by South American rain-forest Indians had the potential to be used as food, medicines, or raw materials for industry. Yet every minute, almost 90 acres of tropical rain forests are destroyed. Since most of the world's species have not even been identified, it is impossible to know what valuable resources are being lost as species become extinct.

Extinctions represent a threat to **genetic diversity** as well. Each species is like a library of genetic information. Genes give organisms the characteristics that allow them to survive in their particular environments. Genes are also the raw material on which evolution works. **Evolution** is the process by which new species of organisms arise from earlier species as the local environment changes over millions of years.

Genetic diversity is important for domestic or native crops and livestock as well. For example, when researchers develop new and improved varieties of crops, they often use wild, or undomesticated, relatives as a source of new traits, or characteristics. Wild plants often have resistance to diseases or insect pests that destroy domestic species. But when a species becomes extinct, the genetic information that it has is lost forever.

Human Causes of Extinction

Extinctions caused by human activities are nothing new. It is believed that human hunters are responsible for the extinction of many of the large animals that lived in North America until 8000 B.C. More than 6,000 years ago, people migrated from the Middle East to Africa and took cattle with them. To provide grazing lands for their livestock, these migrants destroyed trees and encouraged the growth of grasses. The destruction of the original ecosystem caused at least some of the native species to become extinct. A similar process has happened over and over throughout history. Human travelers have brought domesticated plant and animals species with them to new locations around the world. People have also accidentally brought unwanted pests, for example, rats and weeds. By introducing new species, people have disturbed relationships in existing ecosys-

A pronghorn antelope is considered an endangered species. Endangered species are close to extinction.

tems. The new species have no natural enemies and so they often spread rapidly in their new environment. The result in many cases has been the extinction of native organisms.

Hunting and fishing are other ways that people have caused animals to become extinct or nearly extinct. If more animals are killed than can be replaced naturally by reproduction, then a species eventually disappears. Between 1850 and 1920, for example, U.S. settlers hunted so many pronghorn antelopes that the number of animals fell from over 40 million to about 13,000. In Australia, hunters killed millions of Koalas for their skins. By the early 1930s, the species was extinct in South Australia. Today, despite growing public opposition and international treaties, some countries continue to hunt whales. A number of whale species are on the edge of extinction.

Illegal hunting, or poaching, is also a serious threat to animals. African elephants, Indian rhinoceroses, and American black bears are just a few of the endangered species killed illegally each year by poachers.

Wild plants are also at risk because of illegal collecting. The saguaro (sah-GWAH-roh) cactus, a native of the U.S. Southwest, is one example. This species has been seriously reduced in numbers, and its survival is threatened. Saguaros can live to be 150 to 175 years old. They grow very slowly and do not produce seeds until they are at least 50 years old. Healthy plants are sought by collectors who will pay up to $50 a foot for a good specimen.

Pollution

Another threat to species survival is pollution. You have probably seen or heard news reports about oil spills in

The saguaro cactus is the state flower of Arizona. The cactus is now in danger of becoming extinct.

which hundreds or thousands of birds, fish, and marine mammals were killed. **Acid rain,** caused by the burning of oil and coal, is responsible for the death of fish, frogs, and salamanders in lakes in Canada, the United States, and parts of Europe. Acid rain also causes trees in forests to die.

Other pollutants also affect plants and animals. In Great Britain, many bird species are disappearing. Kingfishers are being poisoned by heavy metals that contaminate streams. Skylarks, corn buntings, and linnets are disappearing because the seed-bearing weeds they feed on are being destroyed by chemical **herbicides.**

Hundreds of dolphins in the Mediterranean Sea and thousands of seals in the North Sea have died in recent years because of water pollution. Scientists think that heavy contamination by chemical pollutants weakened the animals' immune systems. As a result, they died from infection with a virus. Throughout much of Europe, otters have disappeared from rivers because of PCB poisoning. PCBs are very toxic chemicals used in cooling systems and electrical transformers. The otters feed on fish contaminated with PCBs, and deadly amounts of the poison are built up in their organs.

Habitat Loss

Loss of **habitat** is the single largest threat to the survival of different species. As the human population grows, people convert more ecosystems to suit human needs. Forests are cut down for lumber and to create open space for agriculture or development. Forests also often sit on land that contains valuable minerals. Mining operations in forests release toxic chemicals that poison water and soil. Rivers are dammed to generate power for hydroelectric plants. The dams then flood huge areas of land, drowning plants and animals alike. Wetlands are filled in for more development projects.

The list of habitat destroyed by human activities is endless. For example, in Texas the golden-cheeked warbler has been officially listed as an **endangered species.** An endangered species is one whose numbers are so low that it is in immediate danger of extinction. This little bird migrates to Central America for the winter and returns to central Texas in the spring to nest. The birds raise their young in mature Ashe juniper woodlands. But huge areas of juniper forests have been destroyed to clear land for farming and expanding cities. As a result, the number of golden-cheeked warblers has dropped to between 4,800 and 16,000 pairs. The U.S. Fish and Wildlife Service estimates that another 50 percent of the population may be lost by the year 2000 if destruction of the warbler's habitat continues.

Saving Species from Extinction

Around the world, people have begun to recognize their responsibility in protecting wild plants and animals from becoming extinct. Many efforts are now underway to save threatened and endangered organisms.

One approach to species protection is to create laws that control activities that might harm wildlife. The U.S. Endangered Species Act, passed in 1973, is one such law. Under this law, organisms listed as endangered are protected from different activities, such as hunting and trapping. Development projects that might reduce or destroy the habitat for endangered species are often limited or prevented completely. However, the law is expensive and difficult to enforce. There have also been many efforts to weaken the law by land developers, mining and logging companies, and others who feel the law interferes too much with their business.

Another important way to protect species is through international treaties and conventions. Many species migrate from one country to another. It is necessary, therefore, to have the cooperation of all the nations in which a species is found. One important international breakthrough was CITES—the Convention on International Trade in Endangered Species. Over 100 countries have adopted this treaty, which outlaws international trade in over 600 of the world's rarest plants and animals. CITES also requires licenses for the export of certain other species. Unfortunately, illegal trade in endangered wildlife is big business. A parrot from the Amazon, for example, can be sold for $5,000. A fur coat made from endangered South American ocelots can be sold for as much as $40,000.

Many organizations can give you information about protecting endangered wildlife. Refer to the Community Resource Directory for more information.

Setting aside land and water to protect habitats is another important method to protect species. In the United States, for example, there are 456 **wildlife refuges.** Other countries have also set aside land for refuges. Thailand recently increased its Khao Ang Ru Nai Wildlife Sanctuary to more than 700 square kilometers. Among the animals protected in this refuge are elephants and tigers. The small Central American country of Belize, recently created its first national park. The protected land is made up of tropical rain forests that are home to such endangered animals as jaguars, keel-billed toucans, and the river otters.

Zoos and botanical gardens work to save endangered species by breeding them in captivity. The intent is to raise organisms that can then be returned to the wild. Unfortunately, not all species will reproduce as captives. In addition, as habitat destruction continues, there may be no place in which to release the protected organisms.

The work of saving Earth's millions of species is enormous and complicated. It requires changes in the way people live all over the world. People in industrialized nations, like the United States, Germany, and Japan, will have to learn to use fewer resources and to use them more wisely. In developing countries, which account for much of the world's population growth, people need ways to support themselves and their families that do not destroy future resources.

DISCOVERY

WHAT ARE HABITAT DESTRUCTION AND EXTINCTION?

Imagine that you are an animal living in a thriving forest community. All of the organisms in the forest live together in a delicate balance. You, for example, depend on many other organisms for food and shelter. Now imagine that a bulldozer is about to destroy a large area of the forest. What will happen to the delicate balance of the forest? Will the survival of some populations be threatened? Try this activity and see.

Materials (per group)

labels   pencil and paper

Procedure

1. In this activity, your classroom will represent a forest community and you will be one of the animals living within the forest. At your teacher's direction, stand at the right side of your desk. Your teacher will hand you a label (face down) designating you as either a "wolf" or a "deer."

2. At your teacher's direction, look at your label. If it says "wolf," you are a predator; if it says "deer," you are the prey.

3. If you are a wolf, stretch out your arms and tap any deer that you can reach. You may not move your feet to tap a deer. If you are a deer, you also may not move unless you are tapped. If you are tapped by a wolf, you have been killed. Remove yourself from the activity and move into the area designated by your teacher.

4. Your teacher will collect the labels from all of the remaining animals and place them in a pile. If many deer have been killed, your teacher may choose to add more "wolf" labels to the pile. This is because populations increase rapidly when food is plentiful. In this case, the wolves have eaten many deer and their population has increased.

5. Now, imagine that half of the forest has been destroyed. Your teacher will designate half of the classroom as the destroyed area and the other half as the remaining forest area.

6. If you are one of the remaining animals, choose a spot in the half of the forest that remains. Once you choose your spot, you must remain where you are.

7. Your teacher will again hand out labels face-down. When you are told to look at your label, repeat Step 3.

8. Your teacher will once again collect the labels, add more "wolf" labels, and designate half of the remaining forest as having been destroyed. If you are one of the remaining animals, find a spot in the quarter of the original forest that still remains.

9. Repeat Steps 5, 6, and 7.

Observations

1. How did the destruction of the forest affect the size of the deer population? Explain your answer.
2. How did the destruction of the forest affect the size of the wolf population? Explain your answer.

Conclusions

1. Predict what eventually happens to the deer and wolf populations. Explain your reasoning.
2. Because this activity focused only on the relationship between wolf and

deer populations, it ignored many factors that would affect this relationship in a real forest. List at least three additional factors that influence the size of wolf and deer populations in a real forest.
3. For each of the factors you listed in question 2, predict how the destruction of a large portion of the forest would affect it. Give reasons to support your predictions.

BRAINSTORMING

Collaborative Learning What are some problems or issues that relate to the extinction of plant and animal species? Spend at least five minutes with a group of students making a list of issues concerning plant and animal extinctions. Express each issue as a question. For example, should much of the world's tropical rain forests be destroyed for economic purposes? Should native peoples be allowed to hunt animal species that are threatened with extinction? Remember, when you brainstorm, do not criticize one another's ideas. After your group has completed the list, have one group member report to the rest of the class. How many issues did the class come up with? Keep a copy of your brainstorming ideas in your notebook. You will need it for some of the activities later in the module. For tips on brainstorming, refer to page 231 of this book.

SCRAPBOOK

Make a Photo Album Make a photo-album scrapbook of species threatened with extinction. Possible sources of such photos and graphics are newspapers, magazines, and pamphlets or brochures. Along with each photo include basic information such as the organism's name, habitat, current and former range or population size, unique characteristics, reasons for its decline, and actions, if any, being taken to preserve it. Continue adding to your scrapbook as long as your class studies plant and animal extinction. For tips on making a scrapbook, refer to the skill lesson on making scrapbooks on page 232 of this book.

JOURNAL WRITING

Critical Thinking In your journal, express your opinions as to why and how humans cause extinctions. Does there have to be a conflict between human need and the needs of other living things? As your class studies plant and animal extinction, continue adding to your journal. For tips on keeping a journal, refer to the skill on keeping a journal on page 233 of this book.

An African elephant is one of the many endangered animals you may include in your scrapbook. Don't forget to add endangered plants as well!

1 FOCUS ON...
Killing Foxes to Preserve Birds

The California clapper rail is a brown and white bird that lives in saltwater swamps and marshes of the San Francisco Bay Area. Fewer than 500 clapper rails remain today, and they are being hunted by red foxes. The short-term question is: Should foxes be killed to help save the clapper rails? The larger question is: Is it appropriate to kill members of one species to help preserve another? The decision has been a difficult one.

BEFORE YOU READ

Decide whether you agree or disagree with the statements below. If you disagree with a statement, change it to make it something with which you could agree. Write your new statements on a piece of paper. Be prepared to defend your choices.

1. People have the right to decide which animals should be killed.
2. It is fair to kill one animal if it endangers another animal.
3. The government should never allow a species to become extinct.
4. If a species is going to become extinct in a short time, the government should not spend large quantities of money to save the species.

WHILE YOU READ

Solving a problem can be difficult. Many solutions have both benefits and costs. Solving one problem may create new problems. Use the chart below to help you analyze the solutions to the problem of the California clapper rail. Copy the chart on a separate piece of paper. For each solution shown, write one benefit and one cost.

Solution	Benefit	Cost
1. Trap foxes and relocate them.		
2. Kill foxes near clapper rails.		
3. Allow clapper rails to become extinct.		

Foxes Hunted to Save Rare Bay Birds

Managers of wildlife refuge take drastic steps against predators

By John Wildermuth
FROM: *THE SAN FRANCISCO CHRONICLE*
MAY 13, 1991

Should the red fox be hunted to save the California Clapper Rail?

Biologists at a bay wildlife refuge have decided that red foxes have to die if an endangered species of bird is to live.

Trapping and shooting of the foxes began this month, as managers of the San Francisco Bay National Wildlife Refuge scrambled to stop the voracious [having a large appetite] predators [animals who feed on other animals] from eating the world's 500 or so remaining California clapper rails into extinction.

"We can't afford to wait," said Rick Coleman, who manages the Fremont-area refuge for the U.S. Fish and Wildlife Service. "The foxes have pups to feed now, and the rails are nesting."

Unlike other breeds of fox, red foxes do not mind the saltwater sloughs [swamps] and marshes where the dingy brown and white rails make their homes. Coleman sees the foxes swimming from island to island, looking for a place where they can ambush birds or steal eggs from nests hidden in the cord grass and pickle weed.

Red foxes are not the only animals willing to make a quick meal of the chicken-sized rails or their eggs. Raccoons, Norwegian rats, feral [wild] cats and striped skunks also hunt the salt marshes that border both ends of the Dumbarton Bridge.

"If (the bay) still had all the tidal marshes it had in 1900, there wouldn't be any problem because the rail nests would be spread over a wide area," Coleman said. "But since most of those

What will happen to the Clapper Rail if it is not saved?

marshes have been converted to salt ponds, highways and industrial parks, it's easy for the predators to find the birds in the tiny remnant left."

When the Fish and Wildlife Service first proposed killing the red foxes in August, it provoked soul-searching among environmental and animal rights groups anxious to save the clapper rails, but unwilling to see foxes die.

"There was some disagreement among us because of the animal rights issue," said Tom Espersen of the Sierra Club. "We finally wrote a letter that was supportive of the effort to save the clapper rails."

Some Bay Area humane associations were concerned that the plan called for using padded leg-hold traps and government hunters to capture and kill the foxes, while animal rights supporters argued that the foxes should be saved.

"We don't support trapping and killing the foxes," said Doll Stanley of In Defense of Animals in San Rafael. "Not enough effort is being made to relocate the foxes that are trapped."

There is really no place to send the foxes, Coleman said. Because the red fox was brought to California in the 1800s by hunters and fur ranchers, it cannot be relocated within the state. And every place else has quite enough of the far-from-endangered animals.

"Every state we've written to about relocating the foxes has said, 'Please no,'" Coleman said.

Catching the foxes is the first problem. Although 10 cage-like traps have been set out in the marshes for a week, a pair of feral cats have been the only animals to stumble into them. Live-trapping adult foxes is difficult, Coleman said, because the animals are smart.

"I'm not hopeful the live traps can make a difference," he said. "More lethal [deadly] means might be needed very soon."

That could include setting the leghold traps or sending out hunters to spotlight the red foxes at night and shoot any of the animals found in the salt marshes, actions likely to bring protests from animal rights advocates.

Some wildlife experts believe that the foxes might be dying in vain. The number of California clapper rails has fallen from 4,000 in 1975 and 1,500 in 1986 to fewer than 500 today. As more of the salt marsh habitat the rails need disappears, the chances the species will survive even the next 10 years are dwindling.

1 GETTING INVOLVED

Critical Thinking

Points of View According to the article, the Sierra Club wrote a letter supporting the effort to save the clapper rail. Write your own letter, using information from the article to back up your opinion. Then write a letter as a member of a group that is opposed to trapping red foxes. State the reasons why the red fox should be allowed to remain in the San Francisco Bay Area.

Gathering Information

Find Out More This article is about the delicate balance in nature and how people can upset that balance. Call or write to the Sierra Club or the World Wildlife Fund (or any other environmental agency listed in the Community Resource Directory) and find out about the balance of nature in your area. What endangered or protected species are there? How have people threatened local plants or animals, and how should these plants or animals be protected? Present your findings in the form of a written report.

Cooperative Learning

Debate the Issue Go to your school or local library and do some background research on the red fox and clapper rail bird. Organize a group of students and hold a debate. One team represents the red fox and the other team represents the clapper rail. Each team should begin with a two-minute opening statement. Then the red fox team asks the clapper rail team three questions, and the clapper rail team asks the red fox team three questions. Finally, each team presents a two-minute closing statement. After the debate, have the class vote to see which team had the most convincing argument.

2 FOCUS ON...
No Room to Roam

When wilderness areas are disturbed so that people can build houses or cities, the habitats of many plants and animals can be threatened. Who decides what should happen to the land—developers who build the houses or conservationists? Is a new community more important than land set aside for wildlife? What are the problems posed when people build housing developments? If an animal is in danger of becoming extinct, should anything be done to prevent this from happening? What can be done?

BEFORE YOU READ

Number a piece of paper from 1 to 7 and copy the statements listed below. Next to each statement on your paper, write **SA** if you strongly agree with the statement, **A** if you agree, **D** if you disagree, and **SD** if you strongly disagree.

1. Any person who builds in the wilderness should help protect the animal and plant habitats of that region.
2. People should pay extra taxes to protect endangered wildlife.
3. The extinction of one plant or animal species has some kind of effect on some other species of plant or animal.
4. People and wild animals cannot live in the same area.
5. The extinction of one species has an effect on urban populations as well as rural populations.
6. Developers have a right to buy and develop land as they see fit as long as they are within the bounds of the law.

WHILE YOU READ

As you read this article, you will find out about several animal species that are threatened in some way. Copy the chart shown into your notebook. Use the chart to help you identify the threat to each animal as well as the effects of that threat.

Animal	Threat to the Animal	Effect of the Threat
light-footed clapper rail		
cougar		
deer		

No Room to Roam

By Marla Cone
FROM: *THE LOS ANGELES TIMES*
DECEMBER 21, 1990

For eight years, the cougar dodged the man-made hazards that have turned the foothills of the Santa Ana Mountains [in California] into a suburban minefield.

In the canyons near Rancho Santa Margarita, she hunted for deer within earshot of bulldozers grading [leveling the land to make] a new road and within driving range of a golf course. At night, she roamed between Mission Viejo and Camp Pendleton along an oak-lined ridge, staying clear of the nuclear power plant and Interstate 5.

Then, on a Wednesday evening in October, her luck ran out.

The 80-pound cougar ventured across [tried to cross] Ortega Highway toward Caspers Regional Park, as she had several times before. But this time she froze in the path of an oncoming car, blinded by the glare of its headlights. She was struck and killed, the latest victim of the urban [city] squeeze that has reduced Orange County's cougar population to less than 20.

Species by species, many of the county's native animals are headed toward extinction because development is slicing the county into isolated patches of open space, federal and state wildlife biologists say.

Many pathways connecting the county's prime wilderness—such as those linking the Santa Ana Mountains with the Chino Hills, or Newport Bay with the San Joaquin Hills—have been blocked, leaving animals with few safe places to roam. Some animals are killed trying to cross traffic, while others gradually are cut off from food, water, shelter and mates. Still others teeter on the edge of extinction because nature's critical balance of predators and prey has been disrupted.

About 90% of Orange County's natural habitats, from oak woodlands in the canyons to salt marshes along the coast, are already gone, and developers and highway builders are planning to dig deeper into prime animal habitat. . . .

Animals will soon be confronted with new obstacles in the county's last frontier—its southern and eastern canyons bordering the Santa Ana Mountains and the coastal canyons between Newport Beach and Laguna Beach.

Three proposed toll roads could turn those areas into a patchwork quilt. Also, three large communities already have been approved—Irvine Coast, Las Flores and East Orange—and several other projects are expected in Coal and Gypsum canyons east of Anaheim Hills. . . .

Developers say they are struggling to balance the needs of nature with the needs of people in Orange County. Inevitably, any development will restrict animals, no matter how carefully planned. But whenever possible, developers and planners say, they are preserving open space and wildlife corridors and building road crossings so animals can migrate.

But wildlife experts say developers aren't doing enough. To compensate [make up] for damaging prime habitat, builders and planners are trying to change the habits and homes of animals, which biologists call a risky game with nature.

At stake is Orange County's rich heritage of wildlife. Unless the pace and pattern of development change, wildlife biologists predict that mountain lions [cougars] will disappear, deer will be driven out in droves, golden eagles will no longer soar over the canyons, and the rare gnatcatcher and cactus wren will be silenced.

The county has 24 threatened or endangered [species of] animals and plants—including two lizards, a fish, a snail and seven birds, and each one is an indicator that an entire ecosystem [community of animals, plants, and bacteria and its environment] is imperiled [in danger]. . . .

The light-footed clapper rail lays her eggs along the banks of Upper Newport Bay and carefully hides her nest under a canopy of cord grass.

But nature's balance has been upset here. And the bird can only watch helplessly as predators such as skunks and foxes ravage her retreat, killing her nestlings [destroy the bird's nest and kill its young].

Because housing tracts and roads crisscross the San Joaquin Hills between Irvine, the Newport Beach shore and Laguna Beach, coyotes have trouble reaching the bay. As a result, the coyotes' prey—red foxes, skunks and cats—multiply to unnatural proportions and threaten the survival of the clapper rail.

Only 131 pairs of the rare birds are left in Newport Bay, and if they are wiped out there, the species will probably become extinct. About 70% of the United States clapper rail population lives in Newport Bay, said Dick Zembal, a biologist with the U.S. Fish and Wildlife Service. . . .

The plight [difficult situation] of the bird is a sign of the ecological trauma [upheaval] unfolding in the coastal canyons of the San Joaquin Hills.

The canyons have about 25 square miles of nearly pristine [unspoiled] habitat—a mix of shrubs, grasses and trees called coastal sage scrub—that feeds and shelters species such as deer, coyote, rabbits, lizards and songbirds.

But wildlife biologists say the habitat in the area is already so fragmented and disturbed that animals have to cross 5 miles of suburban landscape to reach the San Joaquin Hills from the Santa Ana Mountains. The last mountain lion in the area was seen four years ago, and deer and coyotes are dwindling.

The proposed San Joaquin Transportation Corridor could cut off more wildlife migration routes from canyons to the bay, which could finally spell the end of the clapper rails, Zembal said. . . .

Steve Letterly, environmental manager of the county's Transportation Corridor Agencies, said the agency has not decided on the specific design of the toll road, but the agency is committed to building as many animal crossings as possible. . . .

If biologists had a wish list of Orange County habitat they want to protect, most would rank Gypsum and Coal canyons at the top.

Wildlife Corridors

Orange County wildlife is being squeezed by development into ever-shrinking parcels of land. Freeways, roads and housing tracts are fragmenting the migration paths of mountain lions, coyotes and other animals, often cutting them off from food, mates and shelter. The county's largest wilderness areas are in the San Joaquin Hills and the canyons east of Interstate 5, but toll roads and new communities are planned there. Developers and environmentalists are searching for ways to preserve the migration path, known as wildlife corridors. But biologists fear the developments will lead to local extinction of such animals as mountain lions and deer.

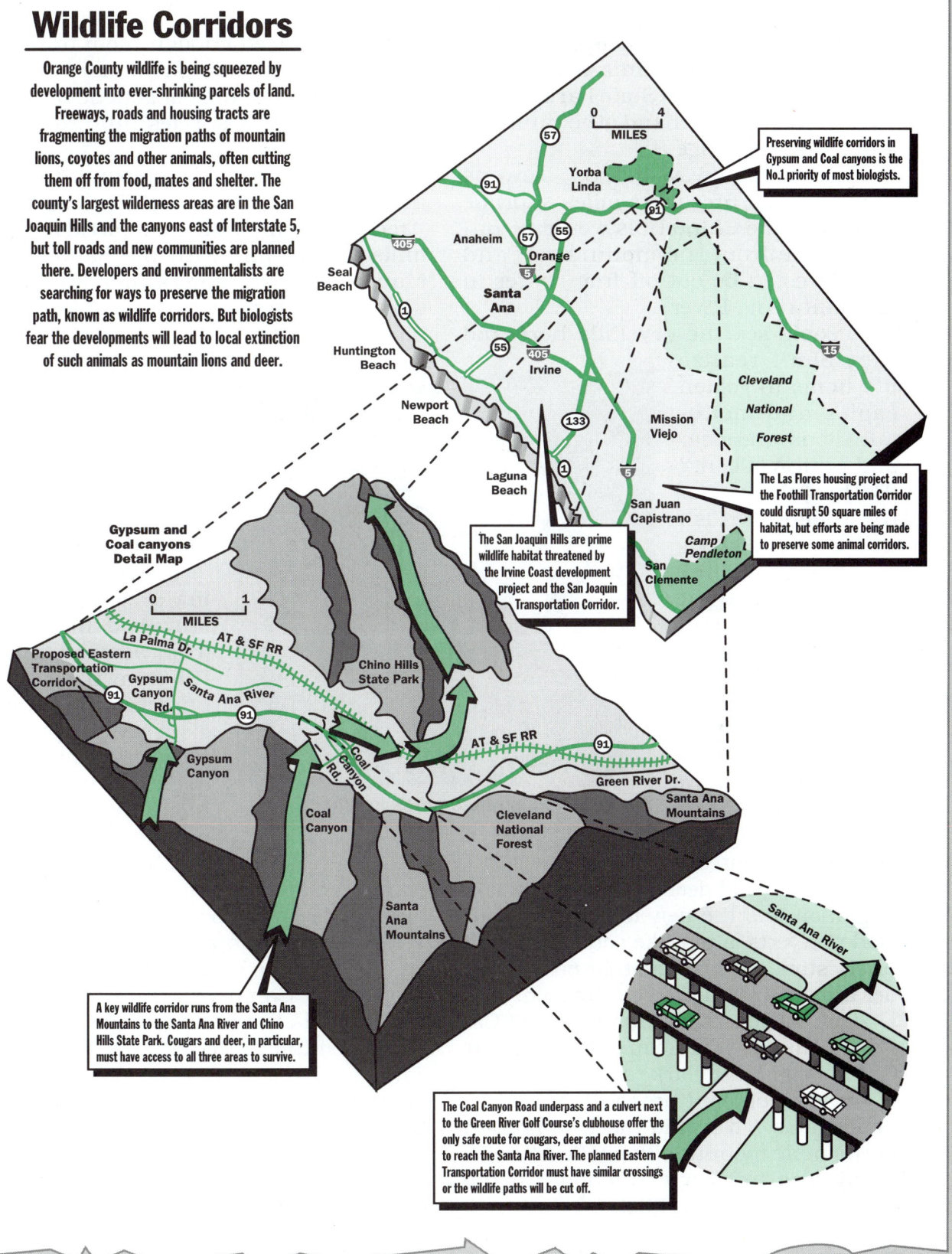

Preserving wildlife corridors in Gypsum and Coal canyons is the No.1 priority of most biologists.

The San Joaquin Hills are prime wildlife habitat threatened by the Irvine Coast development project and the San Joaquin Transportation Corridor.

The Las Flores housing project and the Foothill Transportation Corridor could disrupt 50 square miles of habitat, but efforts are being made to preserve some animal corridors.

A key wildlife corridor runs from the Santa Ana Mountains to the Santa Ana River and Chino Hills State Park. Cougars and deer, in particular, must have access to all three areas to survive.

The Coal Canyon Road underpass and a culvert next to the Green River Golf Course's clubhouse offer the only safe route for cougars, deer and other animals to reach the Santa Ana River. The planned Eastern Transportation Corridor must have similar crossings or the wildlife paths will be cut off.

Populations EXTINCTION OF LIVING THINGS **43**

The side-by-side canyons east of Anaheim Hills link the Santa Ana Mountains with Chino Hills State Park, giving mountain lions, deer and other animals a broad home range.

If animal pathways in those canyons are cut, about 55 square miles of prime habitat in the Chino Hills would be unreachable from the mountains, and wildlife could be cut off from water in the Santa Ana River.

"People see the dry hills here and they don't realize it's home to golden eagles and mountain lions and deer and prairie falcon," [Esther] Burkett of the [California] Fish and Game Department said. "They drive straight through to Disneyland, and it's a shame because it is beautiful."

. . . To determine what natural resources are at stake in the northeastern canyons, county officials have mounted a study to pinpoint deer and mountain lion migration and the new toll road will include culverts [a pipelike construction of brick, stone, or concrete that passes under a road] or bridges, Letterly said.

Jack Bath, a biology professor at Cal Poly Pomona and former biologist at Chino Hills State Park, said the impact of the development proposals would be "enormous." It would block migration of the male mountain lions, and "eventually lead to their extinction" in the canyons and Chino Hills.

Animals need enough room to roam and large enough populations so there is constant exchange of genetic material. Without it, a species inbreeds and eventually dies out.

"Even the Santa Ana Mountains will be significantly changed," Burkett said. "You can't pinch off that corridor that links Chino Hills with the Santa Ana Mountains. We're making islands by developing in pieces and leaving tiny bits of habitat." . . .

The deer needs range areas in order to survive in Orange County.

For mountain lions, life in the remote, rolling hills around Rancho Santa Margarita has taken a dangerous turn in recent years.

Even if the female cougar had survived her trek [trip] across Ortega Highway in October, wildlife biologists say she probably would have been driven out of her home anyway in the next few years.

The spot near Rancho Santa Margarita where she hunted deer will be graded for a new 1,005-acre development project called Las Flores. The 15-mile ridge she used to cross to Camp Pendleton has been identified as the best route for the proposed Foothill Transportation Corridor. And rural roads all over the area are being extended or widened into urban thoroughfares where she would be risking her life each time she crossed.

Earlier this month, the Board of Supervisors approved the Santa Margarita

Company's Las Flores planned community, which will include about 2,500 homes, a school, a commercial center and about 560 acres of open space.

The rugged, rolling terrain where Las Flores will be built is lined with the county's best remaining forests of oak and sycamore, along with coastal sage scrub, a fragrant mix of short, soft shrubs and grasses that grow on hillsides and ridgelines.

The habitat houses a wide variety of animals, from hawks, owls and falcons to deer and reptiles.

The project's environmental impact report noted that it will have "unavoidable adverse impacts" on sensitive habitat, and it will slice through migration routes used by animals such as mountain lions, "contributing significantly to the ongoing regional constriction [restriction] of wildlife movement."

Proposed roads could cut the migration routes of the cougar.

Officials with the Santa Margarita Company said they worked carefully to minimize the ecological impacts. They preserved a major wildlife route that stretches between Chiquita Ridge and O'Neill Regional Park, and to accommodate the protests of biologists and conservationists, they scrapped [gave up] plans to develop 25 acres that fall on another animal route, said Diane Gaynor, spokeswoman for the developer.

"We do understand the need to preserve open space and conserve our resources, while balancing these issues with other important issues such as housing, roads and schools," Gaynor said.

But [Paul] Beier said obstacles posed by the Las Flores community and nearby developments, combined with the new Foothill toll road, spell trouble for cougars. . . .

In all, about 50 square miles of habitat will be disrupted, and several of the 20 or so cougars that remain in the Santa Ana Mountains will probably die, Beier said.

"You can kiss off south Orange County as far as cougar habitat," Beier said. "The toll road will isolate habitat to the west. It's like putting a wall around it."

The toll road agency has not yet selected the exact route of the Foothill Transportation Corridor, and it is analyzing the wildlife migration routes in the area. Letterly said that at least six animal crossings will be built across or under the toll road if Christianitos Canyon is selected.

If the cougars of south county disappear, the species will be gone from almost everywhere in Southern California except the national forests.

"These cougars," Beier said, "are probably the most precarious [endangered] population in the state."

Cougars, which decades ago replaced bears as the top predator in Orange

County, are like the canary in the coal mine. If they die in south county, it means the region's entire ecosystem is unhealthy.

Cougars are the most sensitive to habitat loss since males need at least 100 square miles of home range to breed and feed, Beier said. But they will probably be just the first of many species to dwindle [decrease], including deer, coyotes and raptors. . . .

"We're already seeing animals of all sorts being hit on Ortega Highway," he said. "There are major corridors left, but most have been cut off. You have pieces of habitat here, and pieces there, and it isn't working. Unless we change, the very reason people move here—the allure of Southern California—will be gone."

2 GETTING INVOLVED

Gathering Information

Interview an Expert Contact the director of your state environmental agency. (The phone numbers and addresses for the state environmental agencies can be found in the Community Resource Directory starting on page 244.) Find out as much as you can about wildlife protection programs in your state by asking questions such as the following: What wildlife management programs does the state operate? How much forest reserve is there in the state? Are there any endangered species in the state? If so, what is being done to protect them? Are there any wildlife corridors to protect animals and allow for their movement?

As an alternative, pick a state other than your own. Contact that State's environmental agency and find out the same information. Present the information you obtain in the interview in the form of a written report. Conclude your report by recommending additional ways to protect wildlife in the state you have chosen.

Gathering Information

Create a Database Develop a database of the animals mentioned in the article. For each animal, provide information in each of the following categories: animal name, habitat (use Orange County habitat names), food source, range, mating habits, life span, endangered status, numbers remaining, factors threatening survival, and the reason the animal should be protected. You will need to do library research to find some of the information.

Critical Thinking

Develop a Food Web Using the plants and animals mentioned in this article, develop a food web. Now imagine that a new community is developed and the cougar population in the area has become extinct. How would this affect the rest of the food web? What effects could this have on the environment? What effects could this have on the new community? Summarize your views on the issue in a written report.

3 FOCUS ON...
Status of the Bald Eagle

The bald eagle, symbol of the United States, has been on the endangered species list since 1974. Since then, the number of bald eagles has increased dramatically. Should the status of the bald eagle be changed? Read this article and see what you think.

BEFORE YOU READ

The list below describes some actions that have helped to save the bald eagle from extinction. Number a sheet of paper from 1 to 4. Next to each number, write the way in which that action has helped to save the bald eagle. Work with a partner to develop your answers.

How Have We Helped the Bald Eagle?

1. The pesticide DDT was banned.
2. The bald eagle is the symbol of our country.
3. The bald eagle was placed on the endangered species list.
4. Money has been collected to save the bald eagle.

WHILE YOU READ

Copy the graphic organizer below on a piece of paper. Then complete it by writing the results of each action in the appropriate ovals. After you have finished, write an answer to the following question: Should the status of the bald eagle be changed from endangered to threatened? Why or why not?

Should We Downlist Our National Symbol?

By Carrie Casey
FROM: *AMERICAN FORESTS*
NOVEMBER, 1990

Smack in the middle of campsite No. 62, a pair of bald eagles have staked their claim. Their keen eyes scan Union Valley Reservoir from a 180-foot ponderosa pine.

Of the estimated 80 bald-eagle nesting sites in California, eagles first returned to this one in the Eldorado National Forest in 1986. Each year since —except 1988, when reservoirs were low and food scarce—they have returned and produced a chick. Absent here for over 30 years, the eagles recent return has sparked biologists' hopes that similar revivals will occur in forests across the continent.

The number of breeding pairs today, approximately 2,600 in the lower 48 states, signals a healthy increase from 1960's census of 800 pairs. Nonetheless, human encroachment [the spread of the human population] has taken its toll; experts estimate that just 200 years ago, 50,000 eagles flourished in the U.S. Given those figures alone, bird enthusiasts would gasp in horror. The good news is that eagles are steadily increasing their populations, and have met goals for reclassification downward from "endangered" to "threatened" in four of five regional recovery plans encompassing the 48 contiguous states.

Reasons for the eagle's success are many. Banning the pesticide DDT was a critical factor in halting the bird's decline. Protective management over the last 15 years by federal and state agencies and environmental groups has also helped reverse the trend.

First indications of a connection between eagle decline and DDT use were discovered in the early 1960s. "We were seeing an awful lot of adults, but not many young," said Daniel James, bald-eagle coordinator for the U.S. Fish and Wildlife Service. DDT inhibits [prevents] the transfer of calcium, causing eggshell thinning. Eggs would crack under the weight of nesting birds, and in some cases chicks were laid with no shells at all. The Environmental Protection Agency banned the pesticide in the U.S. in 1973.

In that same year, the Endangered Species Act was passed. Following a survey conducted by the Fish and Wildlife Service in 1974, the bald eagle (*Haliaeetus leucocephalus*) was listed as "endangered" in all states except Washington, Oregon, Minnesota, Wisconsin, and Michigan, where it was listed as "threatened."

Under the act, a cooperative effort involving the Fish and Wildlife Service and the Forest Service is helping to protect nest trees.

"Anytime we find a bald-eagle nest, the objectives for timber harvest in the

Do you think the American Eagle should be downlisted from endangered to threatened? Call your state Environmental Protection Agency and find out the classification of the American Eagle in your state.

area change completely," said Mary Ann Armijo, wildlife biologist for California's Tahoe National Forest. "No longer are we just growing timber—we're growing eagle nesting habitat. We write a silvicultural prescription [forestry plan] for that stand that meets the needs of the eagle."

Long-range planning is a critical element in the eagle's future. Timber stands are being managed with the objective of producing bald-eagle habitat 100-200 years from now.

"Bald eagles are relatively easy to manage in the context of forestry," said Phillip Detrich, Fish and Wildlife Service biologist. "We're not out there harvesting their forage habitat [area where they search for food]. They depend on lakes and wetlands for food, so they're not directly in the path of the use of forest resources, like spotted owls are." . . .

"Eagles nesting or wintering in areas not under federal or state jurisdiction are in the most danger," Dan James told me. "The greatest threats are going to come from the private sector, where the Endangered Species Act has less authority."

Incorporated into the recovery plans is a breeding criterion [standard] of 1.0 young per active pair. This ratio includes successful as well as unsuccessful reproductive results for all pairs in a given area. Of the five regions—Pacific, Southwestern, Northern, Chesapeake Bay, and Southeastern—the Southeastern is the one farthest from the recovery goal of 600 occupied breeding territories. In 1989, 583 had been identified there.

Downlisting in the Pacific Northwest region is being strongly considered since the number of nesting pairs steadily increased between 1985 and 1989. In 1989, nesting territories reached 788, only 12 from the goal of 800 set in 1985.

> **"No longer are we just growing timber—we're growing eagle nesting habitat."**

Zone-by-zone population goals were set in 1985 with data then available. According to Karen Steenhof, research biologist for the Bureau of Land Management in Idaho, attaining goals within individual zones has been slower. Much information has been accumulated in the past five years, and before any final decisions are made, she told me, the Recovery Team needs to "reconvene to fine-tune the goals."

Despite the *Exxon Valdez* oil spill, eagles continue to thrive in Alaska, where the species has never been listed as endangered and now numbers some 30,000. Their abundance is due largely to the rich prey base of salmon and steelhead trout.

The bald eagle shares the endangered list with some 400 other species that equally deserve our support but are much less visible to the public. "We have a couple of snails that people are less enthusiastic about," said William Radke, endangered species program manager for the Bureau of Land Management. "Those snails are no less important, probably, in the grand scheme of the world than bald eagles." But public awareness is important—people need species they can relate to and care about.

The Fish and Wildlife Service has been criticized for exorbitant [excessively high] expenditures for bald-eagle recovery versus dollars for other endangered species. Biologist Dan James told me, "I think it's more appropriate to spend your money where it'll do the most good, and we should be spending our money on those species that are the most critically endangered." However, recognizing the clout of a status symbol, biologist James went on to point out, "A handful of highly visible species like the bald eagle, California condor, manatee, black-footed ferret, and peregrine falcon can carry the day, funding wise, for less visible and perhaps more needy species."

The eagle is a charismatic [appealing] symbol, and people rally to its survival. So, though funding for it appears inordinately [excessively] high, biologists point out that incoming funds filter down and benefit less visible species as well—even snails.

Views about downlisting differ. James, who monitored results of the public comment period that ended last March 30, said, "It's probably fair to say there was more empathy for leaving the bald eagle as an endangered species than for upgrading its status to threatened."

James noted that "threatened" means significantly diminished, but not to the point where a species is likely to become extinct. "That's really where we think the bald eagle is today," he said. "I don't think the bird is likely to become extinct throughout significant portions of its range, but it's not fully recovered yet either."

Even if the Fish and Wildlife Service downlists the eagle to "threatened," the Endangered Species Act would still afford it protection. Following delisting, a species is still monitored for a period of not less than five years. The act provides that if a species shows a significant decline or something catastrophic occurs—such as a hurricane or pesticide threat that wipes out large numbers—the species would be put back on the endangered list right away.

Since the Paleolithic Age [2,000,000 B.C.—8,000 B.C.], when drawings of eagles were first etched on the walls of European caves, these splendid raptors [birds of prey] have stood for strength and freedom and immortality. They may have been given their first reprieve when they were named our national symbol in 1782. If Congress had taken Benjamin Franklin's suggestion and given that honor to the wild turkey, the bald eagle would most likely be extinct in the lower 48 states today. The bird has hardly regained its original numbers, but its populations are building all across the U.S.

"Bald eagles are special," said Frank Isaacs. "People go out of their way to do things for them."

3 GETTING INVOLVED

Critical Thinking

Write a Position Statement A position statement recommends and explains a particular course of action. With a group of students, write a position statement on the following issue: Should an "endangered" species that is no longer endangered, because its numbers have increased significantly, lose some of its protection after considerable effort and money have been spent to prevent its extinction? After your group has completed its position statement, present it to the rest of the class. See if your entire class can reach a position on the question.

Gathering Information

Take a Survey The article states that there are about 400 species on the endangered list in addition to the bald eagle. With a group of your classmates, go to the library and find out what animals are on the list. Choose 15 of them, including the bald eagle. Ask 10 people to rank the 15 animals from 1 (being the animal that should be saved first) to 15 (the animal that should be saved last). Tally your data and record your results on a graph.

Gathering Information

Send for Information Five states—Washington, Oregon, Minnesota, Wisconsin, and Michigan—list the bald eagle as "threatened." Write to the State Fish and Wildlife Agency of each of these states (Addresses can be found in the Community Resource Directory). Ask for information about the state's effort to protect bald eagle nesting sites.

4 FOCUS ON...
Habitat for the Spotted Owl

After much consideration, the U.S. Fish and Wildlife Service has set aside a total of 6.9 million acres of forest land in Oregon, Washington, and California as habitat for the northern spotted owl. Some biologists and environmentalists think that this is not enough land, and that the spotted owl needs more land in order to survive. Representatives of the timber industry, on the other hand, think that too much land was set aside. They say that the economic harm to people who work in the timber industry will outweigh the benefits of saving the spotted owl. Both sides have strong points of view. Read this article to see which point of view makes more sense to you.

BEFORE YOU READ

Listed below are some factors that could be considered in deciding how much land to set aside to preserve an endangered species. Copy the list onto a separate sheet of paper. Then, rank the factors from the one that you think is most important to consider to the factor that is least important to consider. Be prepared to defend your answers.

Factors

1. How many people will lose their jobs?
2. How many animals of that species are left on the earth?
3. How important is the animal to people?
4. How much will it cost to save the animal?
5. Can the animals live in another environment?
6. Are there any animals of this species alive in zoos?

WHILE YOU READ

Copy the graphic organizer shown onto a sheet of paper. As you read the article, fill in the graphic organizer with the players on both sides of the issue. Where do you stand?

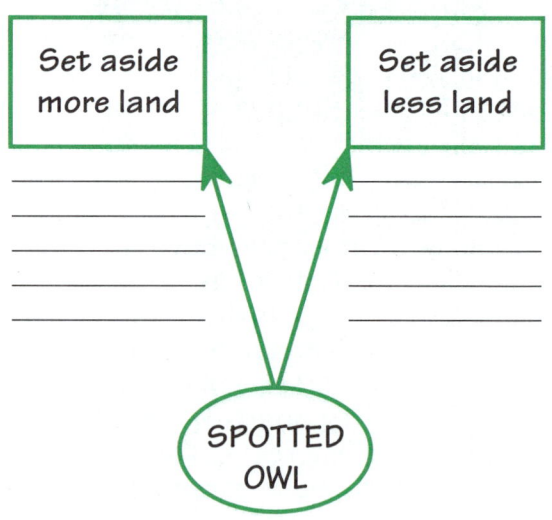

Agency Sets Owl Acreage at 6.9 Million

By Roberta Ulrich
FROM: *THE OREGONIAN*
JANUARY 10, 1992

The U.S. Fish and Wildlife Service on Thursday [January 9, 1992] designated [set aside] 6.9 million acres of Northwest forest land as critical to the survival of the threatened spotted owl.

The timber industry immediately attacked the designation as too much, while environmental groups criticized it as too little.

The Wildlife Service said logging cutbacks resulting from all owl protection measures would cost about 33,000 jobs directly and indirectly tied to the timber industry. About 22,700 of those are direct industry jobs.

But the agency said the critical habitat designation by itself would account for only about 1,420 of the jobs to be lost. And, it added, many of those jobs would disappear anyway if the region's remaining old-growth forests are cut down.

The designation may make little practical difference to the timber industry. The U.S. Forest Service and Bureau of Land Management already were required to consult with the Fish and Wildlife Service before selling timber on federal lands where owls are present.

The habitat designation does not automatically bar logging, but it expands the consultation requirement to lands deemed [considered] essential to recovery of the species even if no owls are present.

The designated habitat for the owl, all federal lands, includes 3.26 million acres in Oregon, 2.22 million acres in Washington and 1.41 million acres in California.

The Fish and Wildlife Service designated 190 areas, mostly scattered along the spine of the Cascades, in the Coast Range of Oregon and on Washington's Olympic Peninsula. The [total] area is 40 percent smaller than the 11.6 million acres the agency first proposed in May, when it included private and state lands as well as the federal forests. No state or private land was included in the new proposal.

> **"I question whether the agency has met the requirements under the Endangered Species Act."**

The final designation also eliminated 869,000 acres of federal lands in counties where the potential loss of logging was expected to have the most serious economic impact. . . .

In an economic summary for the critical habitat plan, the Fish and Wildlife Service estimated that designating the habitat would result in an annual three-state loss of $50 million from the $1.44 billion annual timber value estimated in final long-range national forest management plans. Oregon would suffer the bulk of the loss—$42 million—with Washington losing $5 million and California $3 million.

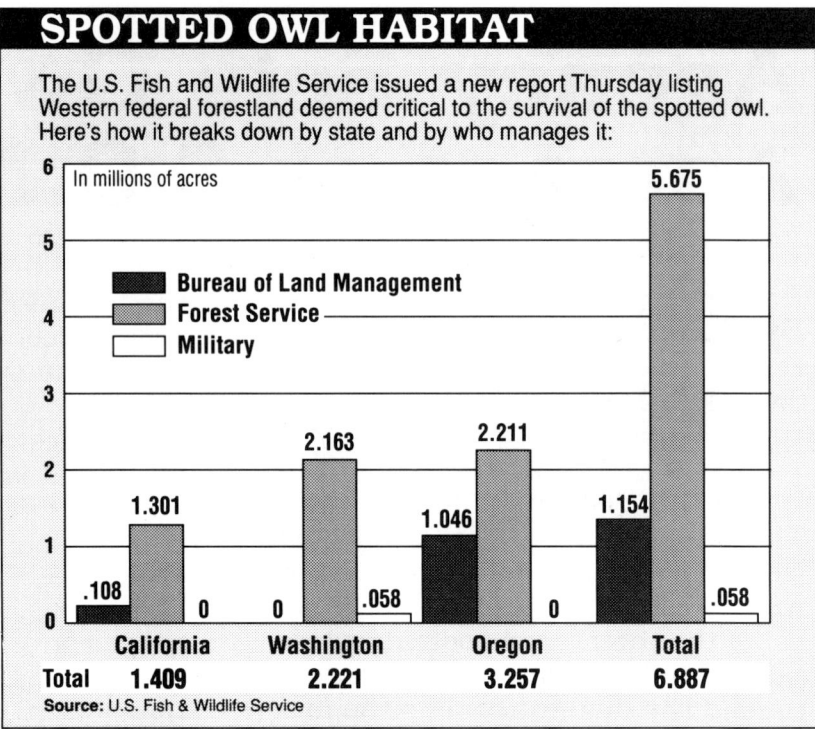

Similarly, Oregon would have the bulk of the job losses: 1,174 out of the 1,420 total. Washington would lose 178 and California 68. The losses are in addition to the estimated 27,705 lost in the three states under the so-called Jack Ward Thomas owl plan and the 3,311 from the 1990 listing of the owl as threatened under the Endangered Species Act.

The report said counties would lose $18.4 million in annual timber sales payments, with Oregon again suffering the major loss: $16.5 million. The federal treasury will lose $44 million annually. Those losses, too, are in addition to losses resulting from earlier owl protections.

"The economic analysis in the critical habitat plan appears woefully inadequate," said Senator Mark I. Hatfield, R-Oregon, "and I question whether the agency has met the requirements under the Endangered Species Act to fully examine the potential impacts of the action."

Valerie Johnson, chairwoman of the Oregon Lands Coalition, a group of people from timber communities, said she was not impressed with the effort to reduce the economic impact of saving the owl. She called the new plan "a half-hearted attempt by the director to throw us a bone. We do not see the Fish and Wildlife Service as an agency that has the slightest interest in minimizing the effects on our people."

Mark Rey, executive director of the industry group American Forest Resource Alliance, said the industry group probably would file a lawsuit challenging the procedures for developing critical habitat. The organization notified the Fish and Wildlife Service in May that it intended to file such a suit but decided to delay until the habitat designation was announced.

David Gardiner, legislative director for the Sierra Club, said he was "troubled by the direction the Fish and Wildlife Service and administration appear to be headed." He said shrinkage of the proposed habitat from 11.6 million acres in May to 8.2 million acres in August to 6.9 million Thursday indicated the timber industry lobby had been very effective. . . .

4 GETTING INVOLVED

Cooperative Learning

What Would They Say? Choose one of the following roles to play:
1. spokesperson for the Oregon timber industry
2. spokesperson for the U.S. Forest Service or Bureau of Land Management
3. spokesperson for the U.S. Fish and Wildlife Service
4. child of a worker in the timber industry who is about to lose his or her job
5. member of Congress who voted for the Endangered Species Act
6. scientist

Have all six people present their points of view to the class. The class then acts as a "town meeting" to see if it can develop a compromise acceptable to all five roles.

Critical Thinking

How Much Area? Find out what the size of your county is in acres. Then calculate how many counties, the size of your county, it would take to equal the 6.9 million acres of spotted owl habitat. Is this a larger or smaller area than you envisioned when you read the article?

Cooperative Learning

Create a Concept Map Work with another student to create a concept map to show what might happen as a result of the critical habitat designation. First, brainstorm to create a list of primary and secondary effects. When creating your list, be sure to include ecological, economic, and social effects. Also indicate what specific groups of people will be affected by the habitat designation. Your map may be similar to the one shown. Draw lines between the concepts that relate to one another. Compare your concept map with the concept map drawn by another pair of students.

5 FOCUS ON...
Preserving the Hawaiian Rain Forest

How can the need for energy be balanced against the need to preserve rain forests? That is the question being asked on the Big Island of Hawaii, where two geothermal power plants have been proposed. The power plants would get their energy from underground steam that is heated by the island's volcanoes. Critics of the plan say that building the power plants will damage the rain forest on the island, home to many endangered species. Read this article, and decide how you feel on this issue.

BEFORE YOU READ

Copy the following energy sources and descriptions onto a separate sheet of paper. Using what you already know about the way energy is produced, try to match each energy source with its description. When you have finished, check your answers using a dictionary.

Energy Sources
1. nuclear
2. hydroelectric
3. geothermal
4. solar

Descriptions
a. energy comes from the earth's heat
b. energy produced from reaction of radioactive material
c. energy produced from the sun
d. energy from water is converted to electricity

WHILE YOU READ

Both sides of the geothermal controversy say that they are trying to help the environment. Copy the graphic organizer below. As you read the article, chart the ways both sides reach the same conclusion. Summarize the main points made by each group on your graphic organizer.

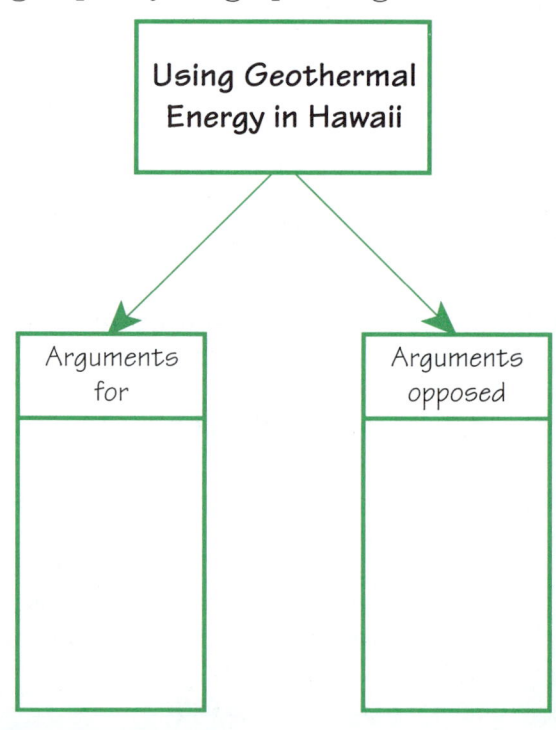

Alternative Energy vs. The Rain Forest

Controversy nearing a climax in Hawaii

By David L. Chandler
FROM: *THE BOSTON GLOBE*
AUGUST 12, 1991

Pahoa, Hawaii—The protesters weave broad green leaves through the wires of an imposing [tall] chain-link fence, covering over a half-dozen "No Trespassing" signs and transforming a steel barrier into a symbolic wall of greenery as they link arms and sing a traditional Hawaiian song. Inside the fence, uniformed guards chase a few infiltrating demonstrators [demonstrators who got inside] away from a huge drill rig.

At first glance, this looks like other confrontations around the world between environmentalists and nuclear power plants or industrial developers. But there's a twist.

Here, both sides earnestly say they are the ones who are trying to save the environment. And they both have a point.

The battle, which may come to a head within a month, pits environmental groups intent on preserving the rain forest against two companies that are trying to build geothermal energy plants to supply the island's electricity.

The environmentalists say the power plants threaten the already fragile future of the Hawaiian rain forest, home to an astonishing variety of plants and animals, most of which exist nowhere else. The companies counter that their impact on the forest will be minimal and that geothermal power is cleaner and safer for the environment than any viable [available] alternative. . . .

Some energy researchers say geothermal [energy] is among the most benign [harmless] sources of electricity, partly because it's so simple.

Other systems use burning fuel or nuclear reactions to heat water, generating steam that then spins a turbine connected to a generator. Geothermal [energy systems] bypasses the middleman. The steam comes straight from the ground, heated by the Earth itself, and into the turbine. After it cools, the water is pumped back into the ground to complete the cycle.

In many places, geothermal energy has racked up a record of reliability, safety, and low pollution levels that is far better than [the records of] most competing sources of electricity. In California, Pacific Gas and Electric gets power equivalent to more than the output of two [large] Seabrook [New Hampshire] nuclear plants from geothermal plants, some of which have operated for 30 years. . . .

Despite the safety records of geothermal plants elsewhere, critics say, the situation here is different: This is the only place where anyone has attempted to build a plant on the flanks [sides] of an active volcano.

The steam is hotter than at other sites—at over 600 degrees Fahrenheit, about 200 degrees hotter than in California. And it may be more unpredictable because of constant movement of the magma, or molten rock, underground.

But while safety is the paramount [main] issue to the residents, the concern that has galvanized interest outside the immediate area is the impact on what some groups tout as [claim to be] the last significant U.S. lowland rain forest—the Wao Kele o Puna.

At 27,000 acres, this forest is a drop in the global bucket: Around the world, more rain forest than that gets cut, slashed and burned every two days. But advocates say this patch is far more

A controversial geothermal drill site in Hawaii.

important than size would suggest because of its staggering biological diversity [variety].

Even though Hawaii takes up less than one-fifth of 1 percent of the U.S. land area, it contains 27 percent of all the nation's endangered species, says Denver Leaman, head of Greenpeace Hawaii. And 72 percent of all the bird and plant species in the nation that have already been driven to extinction were native to Hawaii, he adds.

The extraordinary concentration of unique species on this island—much of it in the rain forest—makes it unusually vulnerable [susceptible] to disruption, Leaman says. Even if the geothermal development takes no more than the 350 acres of rain forest that True Geothermal has permission to clear, that acreage will be largely in the form of roads and power line corridors that will chop up the forest and endanger it all, some biologists say.

"It will become a honeycomb," says Anne Wheelock of the Rainforest Action Group, a leading opponent of geothermal development in Hawaii. "Instead of one large piece of forest, you end up with a whole lot of tiny chunks of forest."

[Allan] Kawada, True's local manager of operations and a lifelong resident of Hawaii, wearily explains that the choice of a site was not theirs to make. Unlike other kinds of generating plants, he says, geothermal plants have to go where the steam is. On Hawaii, the "big island" of the chain, promising sites were identified by the University of Hawaii and the state set aside a geothermal development area that was leased to the firm. The state, Kawada says, carefully considered testimony from biologists about the effects on the rain forest.

"The democratic process made the choice," he says, "we didn't make it."

Some opponents are rigid in their rejection of the technology. The Rain forest Action Group, organizer of many demonstrations, and the Pele Defense Fund, named for the Hawaiian goddess of the volcano—source of all the geothermal steam—oppose any geothermal development in Hawaii, anywhere, anytime. . . .

"We stand in solidarity with the Hawaiian people," says Wheelock of the Action Group. As part of traditional religion and culture, many Hawaiians believe poking holes into the mountain whose volcanism gave birth to these islands is a sacrilege.

And Hawaii is abundantly blessed with other potential clean-energy resources that make geothermal development unnecessary, Wheelock contends. Besides an enormous potential for energy savings through conservation, the islands have plenty of sunshine and steady winds. Wheelock and her husband live what they preach, in a home far from the power lines whose electricity and hot water are provided entirely by the sun. . . .

Other geothermal critics concede the technology might have a place, but only if development proceeds slowly and carefully and is confined to areas outside residential areas and outside the rain forest.

"I think geothermal has a future," Leaman says, "if they go ahead cautiously and check it out carefully as they go."

5 GETTING INVOLVED

Gathering Information

Library Research Visit a library to learn more about geothermal energy. During your research, find out answers to the following questions: How does geothermal energy work? What are the benefits of using geothermal energy? What are the risks associated with geothermal energy? How does the information from your research influence your thinking about what should happen on Hawaii? Organize your findings in the form of a written report that you can share with your classmates.

Cooperative Learning

Hold a Mock Trial Divide your class into three groups. One group acts as lawyers representing the geothermal industry of Hawaii. The second group acts as lawyers representing native Hawaiian people. The third group is the jury. The lawyers for the native Hawaiian people argue the case against drilling geothermal wells as a religious issue, based on the fact that many Hawaiians believe that drilling holes into the mountain is a sacrilege. They claim that the drilling geothermal wells, or any other drilling is a violation of their First Amendment rights to the free practice of their religious beliefs. The lawyers for the geothermal industry develop arguments countering this claim. After both groups of lawyers have stated their arguments, the jury should decide who had the most convincing argument.

Cooperative Learning

Make a Poster Work with three other students. Go to your school or local library to research geothermal energy. You will also have to research Hawaiian culture. Based on the information your group learns, design a poster to show how geothermal energy in Hawaii works. Show the volcano, the steam well, the turbine, and other parts of the power plant. You may want to decorate your poster with drawings or photos that show various aspects of Hawaii and the Hawaiian culture.

6 FOCUS ON...
Preserving Rare Farm Animals

In the past, farmers raised many different breeds of livestock. Today, however, farms around the United States usually raise only breeds that are highly productive. What would happen if a disease wiped out a whole breed of livestock? Is the preservation of livestock breeds important? What consequences could result from the disappearance of some breeds?

BEFORE YOU READ

Create a graphic organizer that shows some important points that you know about farm animals. It may look similar to the one shown below. After you have finished, trade your graphic with a partner and see if each of you can write a short paragraph that interprets the graphic of the other person. You may need to give your partner a hint.

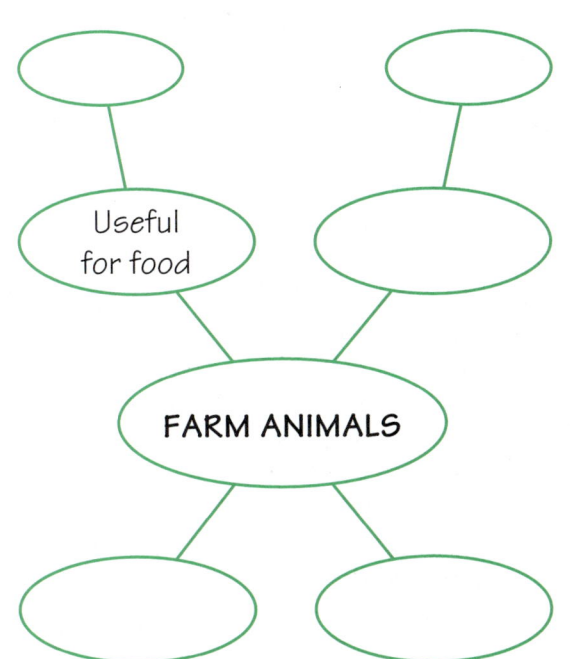

WHILE YOU READ

On a separate sheet of paper, draw a chart like the one shown below. Use the chart to help you decide what, if anything, should be done to help preserve a variety of species of farm animals.

After you have finished reading the article, answer the following question: What should be done about the trend toward raising fewer breeds of farm animals? Be prepared to defend your answer.

Reasons why farmers choose limited numbers of animal breeds	Problems with choosing limited numbers of animal breeds

61

Barnyard Rarities Get Their Day in Sun at Beltsville Show

By Doug Birch
FROM: *BALTIMORE MORNING SUN*
SEPTEMBER 14, 1991

Beltsville—They aren't making farm animals like they used to, and that worries R. John Dawes.

The Baltimore native, who owns [a] farm in Alexandria, Pennsylvania, raises Angus beef cattle for a living. But, like a small number of other farmers, he also keeps a herd of the far more exotic [unusual] Milking Shorthorn cattle, one of about 100 American livestock breeds threatened with extinction.

Mr. Dawes fears that the loss of these endangered breeds would hurt America's rich farm heritage and would mean the disappearance of the barnyard's genetic diversity [variety]. That diversity, he said, might someday be needed to create resistant animals if some virulent [severe and rapidly worsening] disease were to sweep through a popular breed or if climate changes made it difficult for existing animals to thrive on rangeland and farms.

Mr. Dawes justifies the expense of raising the rare dairy cows by implanting Angus embryos in them and using them as surrogate [substitute] mothers for his beef business. But he understands why most farmers in his area concentrate on a few popular types of cattle, chickens, pigs, goats and sheep. Those are the livestock, generally, that produce meat, milk and wool most quickly.

"I think that farmers are more sensitive to environmental issues than you'd give them credit for," he said. "But they're under such enormous economic pressure to meet these quotas, in terms of milk produced and pounds of beef and pounds of wings, they don't have time to address this issue."

In 1976 Mr. Dawes joined the Minor Breeds Conservancy, a North Carolina-based group devoted to preserving endangered farm animals. To help publicize the conservancy's work, he organized an art show that opened yesterday at the National Agricultural Library here [Beltsville, Maryland]. . . .

On a knoll [small hill] outside the library, exhibit organizers set up steel pens and filled them with endangered or just unusual livestock.

A spunky Ossobaw Island Hog—a feral [wild] breed that developed from pigs released by Spanish sailors on an island off Georgia three centuries ago—escaped from her pen and scampered into a copse [dense growth] of pine. It took a posse to corner the panting black-and-white sow and cage her again.

Donald E. Bixby, a veterinarian and

director of the Minor Breeds Conservancy, said that when family farms dominated rural America, farmers raised a wide variety of livestock, many of them hardy breeds well adapted to their environments. Living in the arid [hot and dry] Southwest, Texas Longhorn cattle have adapted to tolerate high levels of calcium in their food, he said. Florida Cracker cattle of the Southeast adapted to the mineral-poor soils there.

Minor breeds were also well suited to the nearly self-sufficient family farm, he said, because they generally don't need grain in their diet, birth easily, nurture [care for] their young and are disease-resistant. But, he conceded, they generally don't grow and reproduce as quickly as more popular breeds.

As factory farms came to dominate American agriculture, economic pressures forced farmers to buy only the most productive breeds. Leghorn hens dominate the egg industry. Holstein cows are so good at producing milk, Dr. Bixby said, "that's practically all we have now" in the dairy industry.

As a result, about half of all American livestock breeds have dwindled to the point where they are in danger of disappearing.

Dr. Bixby said a number of breeds may have vanished without much notice. But the demise of at least two has been documented: the Neapolitan hog and the Lincolnshire Curly Coat pig. The last Curly Coat, he said, was butchered in 1972.

6 GETTING INVOLVED

Decision Making

Take a Stand If a disease or climate change caused extinction of popular breeds of livestock, the supply of milk, meat, and wool could be affected. In a group of five students, discuss the following question: Should U.S. farmers be encouraged and paid by the federal government to breed endangered farm animals? Develop a list of reasons that support your group's answer. Be prepared to defend your answer before the class.

Gathering Information

Find Out More Contact your County Extension Office for information about breeds of livestock raised in your area. Ask questions such as: Are there any rare breeds currently living in your area, or were there any rare breeds in your area in the past? Are any rare or unusual breeds particularly well suited to your area? Present your findings in the form of an oral report to your class.

Critical Thinking

Design an Animal Design your own "rare breed" of farm animal. Sketch what this animal would look like, describe how it would serve humans, habitat, diet, and so on. Why is your creation considered a "rare breed"? Why should it not be allowed to die out?

7 FOCUS ON...
Saving Pandas from Extinction

Giant pandas live in China, where only about 1,000 remain in the wild. Breeding of pandas, in the wild or in captivity, has been largely unsuccessful. The situation is made worse by Chinese government agencies, or bureaucracies, battling each other to control panda research. As you read this article, decide whether you think the panda should be saved from extinction. Is the expense of protecting pandas worth it?

BEFORE YOU READ

Below is a list of reasons why the panda may become extinct. Choose the one that you think will be the main reason for panda extinction and write down all of the reasons for your choice. Then compare your choice with your classmates' choices and see if you agree or disagree.

Possible Reasons for Panda Extinction

1. Bamboo forests, which pandas need for food, are becoming scarce.
2. Female pandas produce few offspring during their lifetime.
3. Panda fur is valuable, and some people are willing to kill pandas to get their fur.
4. Different Chinese government agencies that are supposed to protect the panda are fighting among themselves.

WHILE YOU READ

The chart below is a way of organizing information about panda extinction. Copy the chart on a separate piece of paper. For each problem listed on the left, fill in the chart with the ways that the problem is contributing to panda extinction.

Problem	How problem contributes to panda extinction
Panda digestion is poor.	
There is jealousy among different groups of government officials in China.	
Panda preserves are isolated from each other.	
Panda fur is valuable.	
Pandas are poor breeders.	
Pandas are big attractions in zoos.	

Pessimism Is Growing on Saving Pandas from Extinction

By Sheryl WuDunn
From: *The New York Times*
June 11, 1991

Chengdu, China—In a quiet enclave [small area] in the heart of Sichuan Province, two giant pandas amble playfully among the rocks and bamboo, oblivious [unaware of] to the growing fear that their species could become extinct within decades.

Millions of dollars have been spent to preserve panda habitats and save the animals from hunters who kill them for pelts [skins or furs]. But the efforts have been stymied [stalled because of obstacles] by red tape and infighting among government ministries. The result is that the panda is losing its competition with humans.

"I'm rather pessimistic about the giant panda," said Yin Lin, a technician who conducts artificial insemination [inserting sperm into a female] of pandas at the Chengdu Zoo. "There is a very strong trend toward extinction."

According to China's official statistics, there are only about 1,000 pandas left in the wild, and another 100 in zoos all over the world. Privately, some experts say the number of pandas in the wild is as low as 700.

Loss of Food Supply

Pandas eat bamboo, but their digestion is so poor they must eat practically all day long to get enough nourishment. But in China, where arable land [land suitable for cultivation] and wood are scarce, both local residents and the government have cut down substantial areas of bamboo forests.

Many specialists say the panda's best chances for survival are in the wild, rather than in the zoos, but the wild panda reserves do not necessarily get the funds [they need], and when they do success is limited.

"Our hearts are aching with anxiety," said Pan Wenshi, a panda specialist at Beijing University. "We know the panda must rely on man to survive. But man has not yet offered a good way of helping it."

Researchers and officials are often limited in what they can do: they lack funds, they do not have enough decision-making power, or they are overruled by the bureaucracy.

Sometimes the panda projects are bogged down by petty jealousies and proud bureaucrats, both Chinese and foreign experts say. Money and time are often wasted on repetition of effort, and more money does not necessarily mean better results.

For example, at the Wolong Reserve, a large panda preservation about 90 miles northwest of Chengdu, officials built a research site with $1 million donated a decade ago by the World Wide Fund for Nature. Experts envisioned a program to breed pandas and put them back in the wild. In the past decade, however, the center has bred only one panda, and it died.

One problem is that China's reserves and zoos do not work together to share specialists, resources or even good breeding pandas. Specialists at the Chengdu Zoo have not been welcomed at Wolong, and panda experts from Beijing have been turned away, researchers say.

Part of the reason is that pandas in the wild are controlled by the Ministry of Forestry, while pandas in captivity are managed by the Ministry of Construction.

> **"If poaching isn't checked, the panda will disappear."**

"We do have some contact, but there are some administrative obstacles," says a panda researcher under the Ministry of Construction. "It's a problem. But if we had a leading group overseeing panda work, who would lead it?"

This year Wolong appears to be opening up to outside researchers, and the Chengdu Zoo sent its first male panda to the center for mating this spring. Academics say that while this is progress, there are still many turf wars [disagreements over territory] and problems.

"It's territorial behavior," said Wang Song, a biologist at the Academy of Sciences in Beijing. "The Smithsonian and the American government departments cooperate very well. Under socialism, we should be better but we're not."

China's 14 panda reserves also face other challenges. They are isolated from each other, and sometimes they have so few pandas that the fertile ones have difficulty finding mates. As a result, there are problems with inbreeding as well as relatively low reproduction rates. The World Wide Fund for Nature has proposed linking the reserves together to address these problems, but some experts say this is often not feasible [practical].

Another problem is that even in the reserves, the pandas have two-legged neighbors. Frantic researchers at Qinling Reserve in Shaanxi Province are trying to raise money to relocate the reserve's 2,200 human residents, who have slowly encroached [advanced] upon the bamboo forest.

In Wolong Reserve, new houses have been built for 3,000 residents, but the residents refuse to move unless they are fully compensated for farming income they would lose. Meanwhile, the authorities built a power station to prevent them from further destroying bamboo trees for fuel.

"Everything about pandas gets mired [bogged down] in politics and sometimes it's forgotten that the main purpose is to save the pandas," said George B. Schaller, director for science at Wildlife Conservation International in New York.

Many experts say the panda population has dwindled to 700 to 900 over the past decade, although Chinese officials have been unwilling to announce the results of a panda census completed three years ago, and the official government estimate remains about 1,000. In any case, some experts say the census faced enormous obstacles and was not very thorough. Some say it is possible there are more than 1,000 pandas.

The previous census, in the late 1970's, counted about 2,000 pandas, experts with first-hand knowledge say. But the government announced that the panda population was about 1,000, apparently because it anticipated a decline and did not want to be blamed for it. Some experts believe that the authorities reported a lower number to attract more attention and funds to help the pandas.

Scientists say poaching [illegal hunting] is the principal reason for the population decline. In Wolong, the panda population dropped by half: from 145 to 72 in the 12 years [between 1974 and] 1986, according to a recent report by the World Wide Fund for Nature.

Death Penalty for Poaching

"Hunting is the most serious immediate problem," said Dr. Schaller. "If poaching isn't checked, the panda will disappear."

Pandas have become so prized that the value of a pelt has risen to more than $10,000 to $20,000 in some parts

Millions of dollars have been spent to preserve panda habitats.

of China, according to various Chinese and foreign reports. The pelts make their way through a long chain of underground dealers before they end up abroad, especially in Taiwan and Japan.

China recently decreed the death penalty for poaching and trading in panda pelts, and so far at least four men have been sentenced to death. A recent report in the official Public Security News said that in Sichuan alone there have been arrests in about 200 cases of trafficking in panda pelts in the last few years.

The impact of poaching is particularly serious in light of the panda's poor reproductive abilities. A female panda ovulates once a year, and may raise about half a dozen babies in her lifetime if she is lucky. Breeding in zoos has been difficult because not all researchers know how to tell when a panda is ovulating. Even if a panda is born, it frequently dies in the first few weeks or months, either because the mother accidentally crushes it to death, or because it is not fed properly.

"In the best of worlds, the panda reproduces at an incredibly slow rate," said Devra G. Kleiman, a zoologist at the National Zoological Park of the Smithsonian Institution in Washington. "That's why the loss of even a single breeding period of a single panda is a real loss."

For this reason, some experts say that among pandas in captivity, all efforts should be concentrated on breeding, rather than putting them up for exhibition. But pandas for exhibition have become an enormously lucrative business for the Chinese zoos or institutions that control them.

A Chinese zoo can make more than $500,000 on the loan of a single panda for a few months, and the drive to make money often interferes with panda breeding.

Peter Karsten, director of the Calgary Zoo in Canada, also asserted that Chinese zoos sometimes loaned mismatched panda couples—either pairs of the same sex or pairs that are unlikely to breed—to avoid being embarrassed if a foreign zoo was able to arrange a mating that had eluded the animals' home zoo. He also said that China tries to control the supply of pandas outside China, perhaps for commercial reasons.

A panda must eat all day long to get enough nourishment.

At a broader level, experts complain that there is a fierce competition among institutions that control the pandas, and the result is that zoos and ministries guard, rather than share, pandas as well as knowledge about how to care for them.

Thus, after nearly 20 years of panda breeding in China, only 28 pandas have been bred and raised successfully in captivity there, Chinese panda experts say.

Because of the poor track record, some specialists are skeptical about breeding programs, and more broadly, about panda preservation in China. Even Chinese experts have some difficulty in waxing optimism.

"It will take several generations of researchers before we have success," said Hu Jinchu, a panda expert at the University of Nanchong in Sichuan. "This may be 100 to 200 years. But if the government policy is good, the economy is good, the educational level is improved, and we get international cooperation, then maybe there will be hope."

7 GETTING INVOLVED

Critical Thinking

List the Reasons In at least two places in China, authorities are trying to convince people to move from their homes in order to preserve panda habitat. Work with three other students to create a list of reasons why humans should be relocated to save pandas from extinction. Make a second list of the reasons why the panda should be allowed to become extinct. When your lists are finished, circle the best reason on each list and underline the worst reason on each list.

Critical Thinking

Draw a Conclusion Work with three other students to make a list of the pros and cons of saving pandas from extinction. Develop a numerical scale with numbers from 0 to +5 as advantages to saving pandas from extinction, and from −5 to 0 as disadvantages. Use the scale to assign a number value to each of the advantages and disadvantages on your list. Add up all of the numbers until you get a final score. What is your final score? What conclusions can you draw from your results? Do the advantages outweigh the disadvantages? Report your group's conclusions to the rest of the class.

Problem Solving

Design a Sign Assume that your decision is to protect pandas from extinction. Design a warning sign to be posted near panda habitats. The sign should inform poachers of the value of preserving pandas. It should also explain the penalty for hunting and killing pandas. Within the artistic design of your warning poster, include as much information about pandas as you can.

8 FOCUS ON...
Dehorning Rhinos to Save Them from Poachers

A full-grown rhinoceros may be 6-1/2 feet high and weigh three or four tons. Imagine killing an animal this large only to obtain its horn. That is what is happening in Zimbabwe, in southern Africa, where hunters kill a rhino, cut off its horn, and leave the rest of the rhino behind. Now wildlife rangers are trying to stop the illegal killing by cutting off rhinos' horns. Do you think this is a good way to prevent hunters from killing rhinos?

BEFORE YOU READ

The rhino is prized for its horn. Work with another student to think of other animals that people prize for parts of their bodies. Keep your list on a chart like the one below. Then compare your chart with that of another pair of students, and see how many animals you have in common.

Animals valued by humans	Animal parts valued

WHILE YOU READ

The chart below shows some causes and effects that you will read about in the article. Copy the chart on a separate sheet of paper. As you read, fill in the missing cause or effect for each item.

CAUSE	EFFECT
Rhino horns can cost $12,000 each.	
	Poachers are shot and killed.
Officials are sometimes corrupt.	
Rhinos have their horns cut off.	

Horns of a Dilemma

Zimbabwe is dehorning rhinos in effort to save dwindling herds

By The Associated Press
FROM: *CITIZEN REGISTER* (WESTCHESTER COUNTY, NY)
JANUARY 10, 1992

Harare, Zimbabwe—In a desperate attempt to save the rhino, one of Africa's most endangered species, rangers are cutting off the animals' prized horns with chainsaws to make them undesirable to poachers.

It's not as painful as it sounds, they say.

"Without the horn, they're of no value to poachers," said Dr. Mike Kock, a veterinary surgeon with the National Parks and Wildlife Department. "And since the horn is actually compressed hair-like fibers, it's like cutting fingernails."

The dehorning is a last-ditch bid to protect Zimbabwe's dwindling herds of about 2,000 rhinos in wildlife sanctuaries. The Zimbabwe herds include two-thirds of the world's black rhinos, which are in greater danger of extinction than the white rhinos.

Only 20 years ago, 65,000 black rhinos roamed Africa.

Poachers, often with the help of corrupt officials, have slaughtered tens of thousands of black and white rhinos for horns that are worth about $12,000 each on the black market [illegal trade]. The horns are used for folk medicines in the Far East and as decorative dagger handles in Yemen.

Zimbabwean rangers have killed at least 145 poachers in the past seven years and still have lost at least 960 rhinos to money-hungry poachers.

Despite the risk, the poachers still come, mostly across the Zambezi River from neighboring Zambia in dugout canoes under cover of darkness.

Hornless rhino

In a desperate attempt to save the rhino, rangers are cutting off the animals's prized horns to make them undesirable to poachers.

After horn removal

Poachers kill for horns that are worth about $12,000 each on the black market. Their horns can be as long as 3 1/2 feet (107 centimeters).

- 20 years ago, 65,000 black rhinos roamed Africa.
- About 2,000 rhinos are left in wildlife sanctuaries.
- Zimbabwean rangers have killed at least 145 poachers in the past seven years and still have lost at least 960 rhinos to poachers.

Source: *Associated Press*

8 GETTING INVOLVED

Cooperative Learning

Write a Petition Work with your class and try to come to a group decision on whether rhinos should be dehorned in order to protect them. Then, develop a petition to gain support for your position. Circulate the petition in your school and community. Some students may make posters to put around the school that illustrate the points made in the petition. Try to collect as many signatures as you can. After you have collected signatures, send the petition to the presidents of both Zambia and Zimbabwe by way of their representatives at the United Nations. (The address for the United Nations can be found in the Community Resource Directory.)

Gathering Information

Develop an Action Plan Work with three other students. Go to your school or local library and find out more about this issue. What are the horns being used for? Are rhinos the only animals prized for their horns? What can you do to prevent poachers from killing rhinos and elephants? Develop an action plan that your group or class can follow. Write a brief summary that describes your action plan.

Cooperative Learning

Develop an Editorial Cartoon Get together with three other students and develop a comic strip that relates to the issue brought up in this article. Using the ideas of everyone in the group, create your cartoon. You may need to look in the editorial section of your local newspaper for some guidance. Display your cartoon in your class.

"It's that 'rarer than thou' attitude that gets me."

9 FOCUS ON...
Restricting Trade in Wild Birds

Parrots and other wild birds have become extremely popular pets. Some importers say that bringing the birds to the United States is helping to preserve them because their jungle habitats are being destroyed, and some species may not survive in the wild. Conservationists, on the other hand, argue that allowing trade in wild birds only speeds up the trend toward extinction. This article describes the reasoning behind both points of view and tells what steps are being taken to protect wild birds.

BEFORE YOU READ

Listed below are some words related to this article. With a partner, write three sentences that you think might possibly be in the article. You must use at least two words from the list in each sentence you write, but you may use a word in more than one sentence. Be prepared to give reasons why you think your sentences might be in the article.

Word List

wild birds	campaign	consumer
forest	habitat	pets
inhumane	import	rain forests
parrots	transport	

WHILE YOU READ

The diagram below shows one way of describing both sides of the argument about importing exotic or unusual birds. Copy the chart on a separate piece of paper. As you read the article, complete the chart.

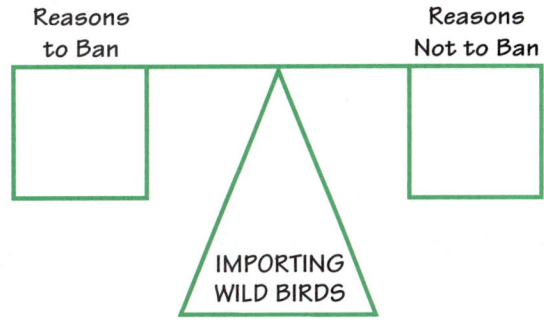

Closing in on Wild Bird Trade

Import ban urged, extinction threat cited

By John Lancaster
FROM: THE WASHINGTON POST
JUNE 11, 1991

At House of Hauser's pet shop in College Park [Maryland], the international bird trade comes home to roost. There amid the cages and cacophony [harsh noises], bird lovers can take their pick of exotic species—parrots, cockatoos, macaws—some of them straight from dwindling tropical rain forests of Africa, Asia and Latin America.

> *"I figure you're saving the species by letting them come into the country."*

All the wild birds can be imported [brought into this country] legally, and owner Chuck Spalding said he thinks he is doing them a favor. "Eventually you won't have any forest, and they're going to become extinct anyway, because of civilization," he said. "I figure you're saving the species by letting them come into the country."

But conservationists contend that just the opposite is true. Alarmed by the threat to native bird populations and by inhumane transport methods that kill 16 percent of wild bird imports destined for the U.S. market, they are pressing airlines, Congress and foreign governments to shut down an industry that they say could wipe out some rare species within decades.

The campaign has gathered steam in recent months, with three air carriers, announcing that they will no longer accept wild birds destined for the pet trade. Honduras last year banned the export of wild birds, and last week two bills were introduced in Congress to prohibit their sale in the United States.

The United States is among the world's largest consumers of wild birds, most of them snatched as youngsters from their nests in the jungles and rain forests of South America, Africa and the Far East. Last year more than 500,000 were legally imported into the United States, and as many as 100,000 more entered the country illegally, according to the U.S. Fish and Wildlife Service.

Experts say the market is growing for both captive-bred and wild birds, particularly parrots, whose iridescent plumage [rainbow-like feathers] and gift for mimicry of humans has endeared them to generations of pet owners. The National Audubon Society warns that international trade threatens to "wipe out" 77 of the world's 335 parrot species.

The bird campaign is part of a growing effort by conservationists to curb the $5-billion-a-year international market in wildlife and products made from it, such as furs and lizard-skin boots. Scientists are concerned that as development consumes rain forests and other endangered habitat, wildlife trappers could push some species into extinction.

"Trade is something we can do something about and some of these habitat problems are more intractable [difficult to manage]," said Susan Lieberman, a biologist who tracks international wildlife trade for the U.S. Fish and Wildlife Service, an arm of the Interior Department. "There are species for whom trade is the number one factor in their decline, as it was for the African elephant." . . .

In theory, [member nations of the Convention on International Trade in Endangered Species] CITES prohibits trade in endangered species and requires countries to monitor other wildlife populations to ensure that trade does not place them in similar jeopardy. But in practice, conservationists say, many developing nations lack the expertise to make such judgments, rubber-stamping export permits on the basis of little or no scientific knowledge.

"In a country like Argentina or Tanzania, a wildlife trader simply goes into an office and asks for a permit," said Ginette Hemley, director of the trade-monitoring arm of the World Wildlife Fund. "There's no science to back it up."

These are some of the Amazon parrots that were confiscated by the Hondurans. The Hondurans have banned the export of these birds.

The persistence of a large legal market also makes life easier for wildlife smugglers, who dodge import prohibitions with phony paperwork and ruses that can include rolling wild birds inside newspapers and stuffing them inside suitcases. Fish and Wildlife officials say they lack the manpower to fully monitor bird imports, inspecting just 25 percent of the shipments that arrive in U.S. ports.

Commercial pressures are particularly acute in the case of parrots and related birds—cockatoos, macaws, Amazons and many other varieties—which account for more than half the wild birds imported into the United States. One rare variety, the black palm cockatoo, can fetch up to $15,000, according to the Fish and Wildlife Service.

To demonstrate the ecological effects of wildlife trading, experts cite the case of the hyacinth macaw, a brilliant blue native of western Brazil that numbered about 100,000 in 1970. The world's largest parrot, the hyacinth macaw was placed on the CITES list of prohibited species, but only after its population dropped to its current level of around 3,000.

Conservationists contend that because 85 percent of the birds sold in U.S. pet stores already are bred in captivity, bird lovers and pet store owners would not suffer any hardship if Congress enacted a ban. Indeed, one bill, introduced by Representative Gerry E. Studds (D-Massachusetts) and Representative Anthony C. Beilenson (D-California), has the support of the Pet Industry Joint Advisory Council.

But that view is not shared by A.A. "Buzz" Pare, one of the nation's largest bird importers, who said an import ban would only hasten the demise of rare parrots whose jungle habitat is disappearing anyway.

"What we're doing is salvaging the birds," he said in a telephone interview from his import business in Miami. "We worm them, delouse them, and find them a home. If they stay in the jungle, they're not going to have it in the next five years. . . . You can see them burning the forests down."

Groups such as the Humane Society of the United States have pointed out what they contend is cruel treatment of captured birds. According to the Fish and Wildlife Service, 16 percent of all wild birds shipped legally to the United States die of suffocation or disease while in transit or during the required 30-day quarantine [isolation] period.

The Humane Society estimates that the true mortality [death] rate is much higher, saying many birds die when they are trapped or smuggled across the U.S.-Mexican border, sometimes drugged and with their beaks taped shut. Negative publicity about high death rates led to the suspension of shipments by the three carriers, which accounted for 44 percent of U.S. wild bird imports in 1989, according to the environmental group Defenders of Wildlife.

Pare, the importer, lost 19 percent of his birds before they left quarantine in

> **". . . once the birds get in an airplane and under stress, they die."**

1985, according to one study. But he said the figure was skewed [distorted] because it included finches shipped from Senegal and other African countries. These birds often have been exposed to pesticides, he said, and "once they get in an airplane and under stress, they die."

9 GETTING INVOLVED

Cooperative Learning

Create a Concept Map Work with another student to create a concept map to show what might happen if wild birds continue to be imported. First, brainstorm to create a list of primary and secondary effects. When creating your list, be sure to include ecological, economic, and social effects. Also indicate what specific groups of people will be affected by the wild bird imports. Your map may be similar to the one shown. Draw lines between the concepts that relate to one another. Compare your concept map with the map drawn by another pair of students.

Critical Thinking

Write a Proposal Work with three other students. Imagine your group is part of the Convention on International Trade in Endangered Species (CITES). You have gathered to discuss a worldwide ban on transporting wild birds. Your group should prepare a proposal to be ratified by the member countries. Include in your proposal the limits on imports and ways to enforce import rules, as well as the punishment that will occur if transportation occurs in defiance of the ban.

Problem Solving

Design a Bumper Sticker Design a bumper sticker that encourages the protection of wild birds in the rain forest. The message on the bumper sticker must be brief and easy to read.

Extinction of Living Things

THINKING CRITICALLY ABOUT LOCAL ISSUES

You may want to work with a group of students for this activity. If so, make sure that each member of the group agrees about the issue that is chosen.

Prioritize the Issues Make a list of issues related to extinction that are important to you and your community. Start by reviewing the brainstorming list that you made at the beginning of this module. If you made a scrapbook or wrote in a journal, look over those as well. Also review the magazine and newspaper articles in this module. Evaluate all the issues on your list. Then, put them in order so the issue that has the highest priority for you is at the top of the list.

Make the Issue Your Own Become an expert on the issue you ranked at the top of your list. Then, decide what you as a citizen can do about this issue. Apply the process that you learned in the *S-T-S Problem Solving* module. Remember, the skills in the process are:

- Analyzing the Issue
- Gathering Information
- Making a Decision
- Planning Action

If you are working in a group, each group member has to contribute at all four stages. After completing the process, share your experience with your class.

CREATIVE THINKING

Write a dialogue that you could imagine taking place between a conservationist and a rain forest landowner. The landowner wants to clear the area to grow crops and raise cattle. The conservationist wants to protect the land. In the dialogue, each person tries to justify to the other his or her position. Try to construct the dialogue so it expresses the values of both sides of the issue.

Wrap-Up

HELP WANTED

A Recruitment Fair With a group of your classmates, make a list of careers that have some connection with the issue of extinction. Some careers to consider are plant geneticist, animal geneticist, wildlife biologist, game warden, zoo manager, environmental lobbyist, and conservationist. You might also check the articles in this module to see if other related careers are mentioned. Select one of the careers on the list and find out all you can about that career. Then, based on this information, make a recruitment speech to your group in which you try to persuade people to pursue the career. Be as positive about the career as you can, but be realistic. After all the recruitment speeches have been made, discuss as a group the pros and cons of each career. Take a vote to determine the most popular career. Have a spokesperson report your group's findings to the class.

A zoologist is one of the many careers related to protecting endangered species.

HUMAN Populations

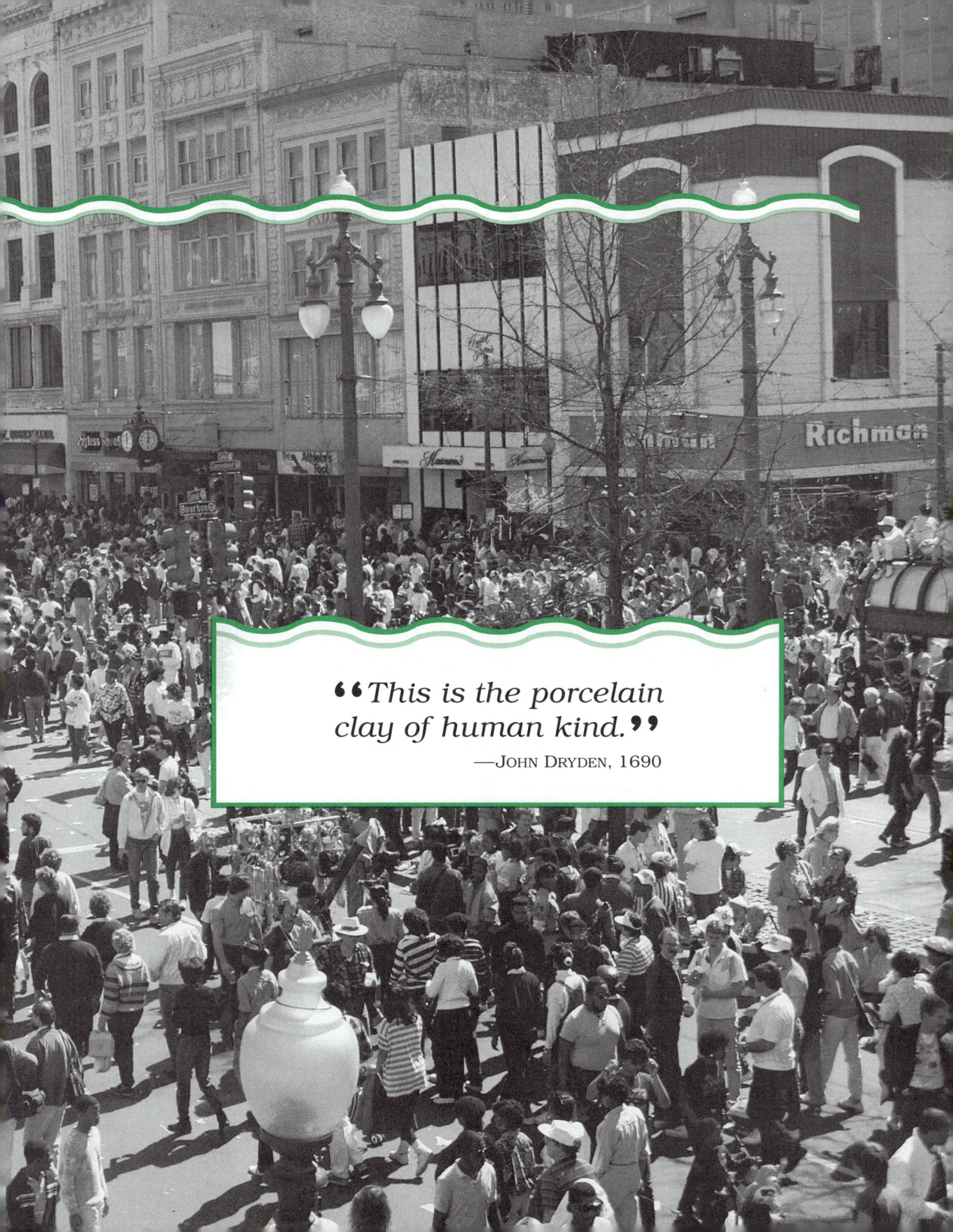

> "This is the porcelain clay of human kind."
> —John Dryden, 1690

HUMAN Populations

IN THIS MODULE

Overview	83
Discovery	90
1 "Groups Unite to Point Out Hazards of Overpopulation" from *The Los Angeles Times*	92
2 "Running Out of Room" from *The Toronto Star*	95
3 "Busting The Boom: Population Control Works" from *WorldPaper*	100
4 "Norplant Renews Debate Over Forced Contraception Reproductive Rights" from *The Morning Call*	105
"Birth Control or Woman Control?" from *Charlotte Observer*	109
5 "Abortion Issue Divides Advocates for Disabled" from *The New York Times*	111
6 "U.S. Laws Under Attack" from *The San Francisco Chronicle*	115
7 "An Investment in American Citizenship" from *The Washington Post*	119
8 "Administration on Aging Announces The National Eldercare Campaign" from *Aging*	123
Wrap-Up	126

TOO MANY PEOPLE, TOO LITTLE SPACE

You're giving a party at your house for 15 friends. At least that's the number of people you invited. You have enough food and drink for that many. Your house can hold 15 comfortably.

The problem is, all of your invited guests have decided to bring along others. More than 30 people show up at your house. Thirty people are now sharing the food and drink meant for 15. There's not enough to go around. So everyone is left hungry. There's no room to dance or have a good time. People are bumping into things. Your mom's favorite vase is in pieces on the floor. You've got yourself an overpopulation problem.

Of course, this is overpopulation on a really tiny scale. But you can see the point. Too many people sharing the same resources means no one gets enough—not enough food, drink, or space in this case. But, at a party, you can always run out and buy more chips and soda. Maybe you can glue together the vase. If it's a nice day, you can ease the overcrowding by moving outdoors.

When you consider overcrowding on Earth, the situation is not so simple. Here on Earth we have more than 5 billion people. We all share a limited supply of food, water, and space. If our "party" gets too crowded, we can't send out for more food or water. Our planet has a limited ability to provide both. We can't move extra people elsewhere. There aren't too many takers for a one-way trip to Mars. That's why it is so important for us to be careful about the number of people that come to our global party—the number of people born on our planet. Because having enough food, drink, and space for everyone is a matter of life and death for us all.

Bangladesh has more than 2,000 people per square mile.

Growing By Leaps and Bounds

More than 5 billion people and counting. That's our planet's human population, and we're still packing them in fast. Since you began reading this article, several hundred more people have been born. The earth adds more than 10,000 people each hour; that's 254,000 each day. Every month we have 8 million more people on Earth than we had the month before. That's like adding 12 new cities the size of New York City each year. Yes, the world's population is really growing. But it has not always climbed this fast.

From the time that people started keeping records a few thousand years ago, until the early 1800s, the world's population grew slowly. There were several reasons for this. For most of human history, for example, people died fairly young. Many women died in childbirth, and many children were killed by diseases before their fifth birthday.

For these reasons, it took tens of thousands of years for Earth's population to reach 1 billion. That happened in about 1800. Then along came small improvements in areas such as medicine and farming. More people survived common diseases. Famine was less widespread, so fewer people died of starvation. The world's population doubled to 2 billion in just a little more than 100 years. Science and technology have improved still more in the 20th century. Antibiotics, such as penicillin, were developed in the 1940s. Later, the **Green Revolution** in farming increased the total amount of food each parcel of land could grow.

It took less than 50 years for the world's population to double again, to 4 billion. That happened in 1974. Just 13 years later, in 1987, the population topped the 5 billion mark.

In contrast to Bangladesh, Canada has only 6 people per square mile.

By the end of the next century, the earth will probably have between 8 and 14 billion people. The earth's land and resources are already starting to show the strain with our present 5 billion. Some scientists think our population is already too large. What kind of world will we have with twice our present number of mouths to feed? Will there be enough homes and jobs for everyone? Will there be enough fuel and clean water? How much more garbage and pollution would these 3 to 9 billion more people create? Can our planet absorb it all? Just how many more people can we safely invite to the party before there are just too many?

How Many Children Is Too Many?

The rate of growth of the earth's population depends on two factors: birth rate and death rate. Birth rate is the number of yearly births for each 1,000 people in the population. Death rate is the number of deaths for each 1,000 people in the population. If death rate exceeds birth rate, the population decreases. If the two match, the population doesn't change. But when birth rate exceeds death rate, population increases. That is our situation today.

Right now, the earth has a growth rate of 1.7 percent. That doesn't sound like much. But it's enough to add perhaps another 2½ billion people in less than 40 years.

One problem is that these people are not evenly distributed on Earth's surface. Growth rates also differ from place to place. About ½ of the world's people live on only 5 percent of the planet's land surface. Population density also varies. That means there are few places with lots of people packed in tightly and many places with plenty of elbow room.

For example, Canada has just 6 people per square mile. Bangladesh has more than 2,000 people per square mile.

Unfortunately, many—though not all—of the countries with the highest growth rates, like Bangladesh, are also those with great population densities. So adding more people just makes a crowded condition worse.

Growth Rate

In general, the growth rates of the more developed countries of Europe and North America are lower than the growth rates of the developing countries of Asia, Africa, and Latin America. There are several reasons for this difference. For example, people in more developed nations generally have higher living standards and higher incomes. They are also less tied to manual labor

People are living longer and healthier lives as technology improves.

on the land, so they see less need to have big families. In developing countries, people have lower living standards, less money, and less formal education. Children are needed to work in the fields to help the family earn a living. Often, children are seen by parents as a way to survive old age in places where there are no pensions or Social Security checks. More children also die in early childhood in some developing countries. So families feel they need to have many children to produce the large families they need.

As a result, there is more suffering in many developing nations. Often these countries already have shortages of food, clean drinking water, housing, medical care, and jobs. Modern medicine has lowered the death rate in some of these places. But modern technology has not been able to keep up in other areas such as jobs, providing shelter, and the distribution of food and drinking water. So the addition of more people puts more stress on local land and resources. It also puts stress on world resources. For example, an overabundance of people in Brazil might force more farmers to cut down rain forests in order to make use of the land. That affects the whole planet, as the lack of trees helps increase carbon dioxide in the air and hastens the extinction of important plant and animal species.

Limiting Birth Rates

Some nations with high growth rates are now trying to limit their birth rates. There have already been decreases in many countries, including China, Sri Lanka, Mexico, Cuba, and Singapore.

China, with 1 billion people, has one-fifth of the world's people. However, China has just 7 percent of the world's land, and much of this is rugged mountains. A few years ago, Chinese officials looked at population projections for the year 2000 and beyond. They saw that at their growth rate, they might have 3 billion people to care for by the year 2030. They then decided to make a plan to reduce their nation's rate of growth.

As a result, China started its one child per couple policy in 1978. Officials used public education, family planning workers, and economic pressure to encourage Chinese couples to have only one child. Couples who had more than one child were even fined. Many Chinese couples then started to have fewer children. China's birth rate has dropped below what it was when the program began. It is now 1.4 percent. This is below the world average. China's population is still growing because there are so many people of child-bearing age. However, China is not growing as fast as it would have without its one-child policy.

Singapore is a tiny city-state in Southeast Asia. Unlike China, Singapore is a prosperous and modern nation. However, Singapore has a population density of nearly 12,000 per square mile—almost the densest in the world. Singapore has no space to add new people. So Singapore put a "two children only" plan into effect several years ago. Small families were able to attend better schools and get better medical care. They even got tax deductions for not having children. Singapore's birth rate also dropped as a result. Singapore is still growing, but very slowly.

Still, there are many poor countries in which the birth rate remains high. The average growth rate in Africa is 3 percent. This is almost twice the world average. In Asia and Latin America, it is about 2 percent—this is still higher than the average of 1.7%. In some cases,

In the United States, the population increases about 1% each year due to immigration.

cultural or religious beliefs prevent the use of birth control. In others, governments are too poor to provide to families the needed education about controlling population size. So, population continues to increase, putting a strain on our scarce resources.

These questions must also be asked. Is it right for governments to control population size? Should families be fined or punished in some way for having more than one or two children? Who can make such a decision? Should rich families be allowed more children than poor families? Is that fair? These are difficult questions and not everyone agrees on the answers to them.

Controlling the rate of population growth can solve some problems, but it can also create others. As population growth rates decrease, the average age of the population could increase. In time this could mean that there would be as many retired people as there are people working. This could create economic problems. Governments might not have enough money to support the aging population. Also, there might not be enough workers to support the economy. This could be especially true in developed nations where advances in medical technology are enabling people to live longer. How do you think these potential problems should be dealt with?

The Environment at The Breaking Point

Many of Earth's 5 billion people are already undernourished, diseased, and living in poverty. At least another 3 billion people will be added over the next few decades. Ninety percent of them will be born in the world's poorest nations. This brings to mind one very important question. How will we feed, clothe, and house all of those additional people when we cannot provide for all of those we have now?

Earth's population has a direct effect on the environment. More people mean greater energy use. More people also mean more pollution. More land is needed for farms and homes. This results in more deforestation. With less resources to go around, there are more conflicts over the available resources between nations or groups within nations. More human beings also means more destruction of other animal and plant species as we destroy their habitats and squeeze them out.

Some experts believe that reducing population is the total answer. But others disagree. They recognize that the earth has limits on how much food, water, fuel, and shelter it can provide. But they don't feel we have reached the breaking point yet—or need to in the near future. They feel that the problem is the uneven distribution of people and resources. What do you think?

All societies have an impact on the planet. We cut down trees and drill wells for oil. We grow food by using chemicals that go into the soil. We create garbage and pollution that spills out into the air and water. A modern society like ours in the United States, increases the impact that each one of us has on the planet. Someone who lives in a large, modern city in the United States uses more food, fuel, water, and other resources than a primitive farmer in the Sudan or a herdsman in the outback of Australia.

Take the United States, for example. People in the United States make up just 5 percent of the world's population, but they use about 25 percent of the world's energy resources and materials. People in the United States produce about 25 percent of the world's pollution, and each one person has the impact on the environment of many people from a developing country.

Regardless of our growth rate, we will reach the planet's limits at some point. Rapid destruction of resources such as forests and farmland, as well as pollution of water and air, result from population increases. How long will it be before natural systems are damaged beyond repair? Along with concern about population size, concern about the environment and conservation of resources will be helpful as well. The less we use and throw away, the less impact we will have on our surrounding environment.

What About The United States?

The United States has many of its own population problems to solve. This population grows by more than 2 million each year. Because of the **baby boom** (a large number of births after World War II), there are a great number of women of child-bearing age right now in the United States. However, the United States birth rate barely exceeds the death rate. Our natural population increase due to births is really not very high. Still the United States population increase nearly 1 percent each year. How? Mostly through **immigration.**

The United States already has problems of pollution, unemployment, hunger, and homelessness. Immigration of large groups of people may make these

In 1886, the French government gave the Statue of Liberty to the United States. This was the first landmark many immigrants saw as they entered the United States.

problems harder to solve. Some people believe the United States should halt immigration or limit it to individuals who could contribute to its wealth. Others believe the United States has an obligation to people who cannot make a living in their native country, or whose civil rights are being violated.

What About Technology?

Technology has also provided means of sustaining larger populations. One example is improved crop yields. Technology has also raised new questions regarding populations. Technology is now available that can detect abnormalities in a fetus. For example, there are two tests, an amniocentesis and a chronic villus biopsy, that can be performed on pregnant women to test their embryo for genetic defects. These new technologies have raised questions about whether such pregnancies should be terminated. Some people believe this would relieve some population burdens. They feel that if resources were not strained by supporting severely ill or impaired individuals, the quality of life would be better for others. Some countries even feel that governments have the right to make these choices. Others feel that these are personal choices and still others think that this is not a choice at all. Who do you think has the right or power to make these choices?

What Do You Think?

Think about the many questions that relate to populations—their size, their distribution, their well-being, and so on. Think about how you would go about trying to answer some of these questions. Would these questions be easy to answer? What kinds of information would you need? Where would you get the information you need?

The articles you are about to read will address some population issues. As you read, consider your response to these dilemmas. What kind of decisions will you need to make?

DISCOVERY

WHAT IS MEANT BY THE POPULATION EXPLOSION?

Did you know that the world's population increases by about a million people every 5 days? If such rapid growth continues, the world's population will double—from 5 billion to 10 billion people—in just 40 years! Can the earth support so many people? Will there be enough housing, food, and other resources for everyone? Or, is there a limit to how many people Earth can support? Try the following activity to see.

Materials (per group)

3 large bowls
water
stirrer
wax pencil
set of measuring cups
food coloring
newspaper
adhesive labels

Procedure

1. Working with a partner, spread a few layers of newspaper around your work surface.
2. Fill two large bowls about three quarters full with water. Add a few drops of food coloring to each bowl and stir.
3. Label one of the bowls "Earth" with an adhesive label, and place it in the center of your work surface. Mark the level of water in the bowls with a wax pencil. The volume of water in the bowl represents the current world population.
4. Label the other filled bowl "Births" and place it on one side of the "Earth" bowl. One partner should stand near the bowl.
5. Label the empty bowl "Deaths" and place it on the other side of the "Earth" bowl. The other partner should stand near this bowl.
6. The person standing near the "Births" bowl should take a one-cup measuring cup. The person near the "Deaths" bowl should take a quarter-cup measuring cup.
7. You and your partner should take turns adding and removing water from the "Earth" bowl. The person near the "Births" bowl should begin by filling his or her measuring cup with water from the "Births" bowl and transfer it into the "Earth" bowl. Next, the person near the "Deaths" bowl should fill his or her measuring cup with water from the "Earth" bowl and transfer it into the "Deaths" bowl. (See the set-up in the figure below.)
8. Continue taking turns transferring water in this way until your teacher tells you to stop. Observe any change in water level in the "Earth" bowl.

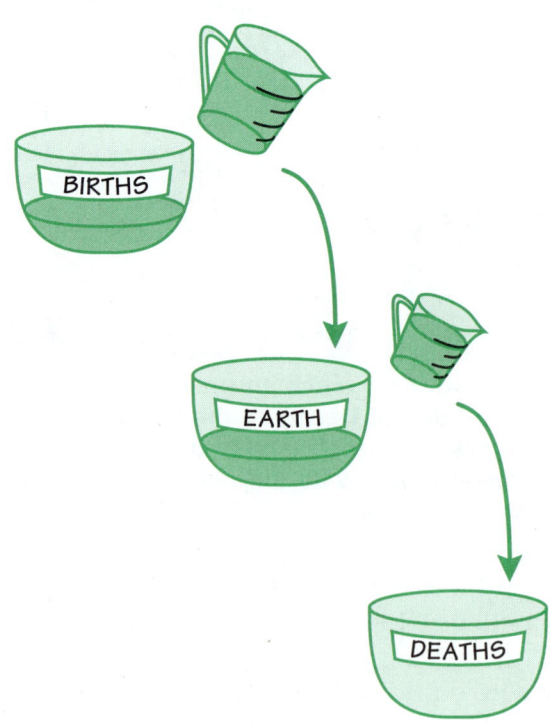

Observations

1. What was the significance of adding water to and removing water from the "Earth" bowl?
2. How did the level of water in the "Earth" bowl change? What does this symbolize in terms of the world's population?
3. Explain the change in water level as a model of the effects of birth rate and death rate.

Conclusions

1. What do you think will happen in the world if the birth rate continues to be greater than the death rate? What do you think should be done to address this problem?
2. List some world problems that you think are either directly or indirectly related to population growth.
3. The term *zero population growth* is used to describe a condition in which the total population remains unchanged. This occurs when the birth rate is equal to the death rate. How could you modify this activity to model zero population growth?

BRAINSTORMING

Collaborative Learning What are some problems or issues that relate to human population growth? Spend at least five minutes with a group of your classmates making a list of issues or actions related to human population growth. Express each issue or action as a question. For example, should governments provide rewards to couples who have fewer children? Should countries without an overpopulation problem be required to accept immigrants from overcrowded countries? Remember, when you brainstorm, do not criticize one another's ideas. After your group has completed its list, have one group member report to the class. How many issues did the class come up with? Keep a copy of your brainstorming list in your notebook. You will need it for some of the activities later in the module. Guidelines for brainstorming can be found on page 231.

SCRAPBOOK

Gathering Information Human population growth directly influences many world events. One way to get a sense of this is to collect newspaper and news magazine articles on events caused by or related to human population growth. For example, food shortages and famines, epidemics, border disputes, emigration and immigration, pollution, and destruction of natural habitats are some examples of events that can result from human population growth. Look for articles on events that are population-growth related. Clip them and place them in your scrapbook. Include an explanation of the relationship between the event and human population growth. Continue adding to your scrapbook as long as your class studies population growth. For some ideas on keeping a scrapbook, see page 232.

JOURNAL WRITING

Critical Thinking Some people claim that overpopulation is the basic cause of all environmental problems. In your journal, write your response to this opinion, including your reasons for agreeing or disagreeing. As you work on the module and your awareness of population-growth issues increases, add to your journal. See page 233 if you need help on journal writing.

1 FOCUS ON...
Dangers of Overpopulation

Trying to solve a major problem in the United States, such as population growth, is made more difficult by the fact that we are a diverse society with many different points of view. Now participants from many groups have come together to deal with the problem of overpopulation. Read the article to find out who they are and what they hope to accomplish.

BEFORE YOU READ

Below is a list of words from the article. Look at the words and write at least three sentences that predict what you think the article will be about. After you finish reading the article, check your predictions.

Words

overpopulation   politics
environment   united front
resources   ecology
coalition   public awareness
family planning

WHILE YOU READ

The chart below shows one way of organizing information about overpopulation. Make a chart like the one shown and fill it in as you read the article. Look for answers to these questions: **Who** are the people concerned about overpopulation in the United States? **What** is their concern? **Why** are they concerned? Do they have any **solutions?**

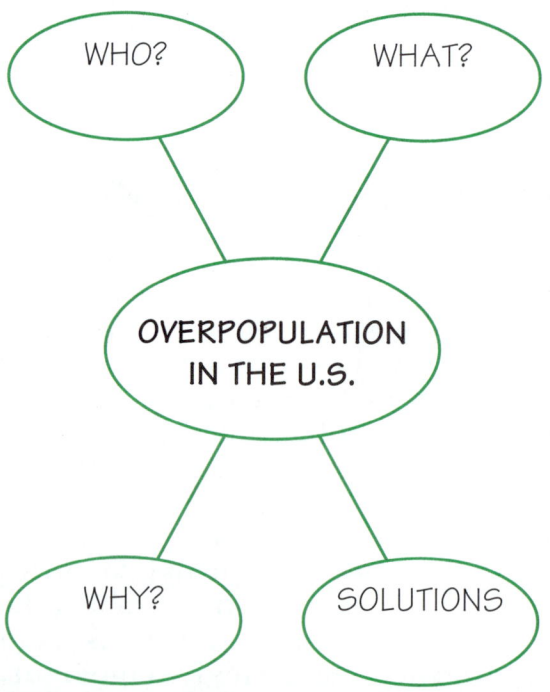

Groups Unite to Point out Hazards of Overpopulation

by Holly K. Hacker
FROM: *THE LOS ANGELES TIMES*
MAY 23, 1991

A diverse group of environmentalists, scientists, Nobel laureates [winners] and others launched a concerted [combined] effort Wednesday to heighten public awareness of the disastrous social and environmental effects of overpopulation. . . .

"Together, the increase in population and in resource consumption are basic causes of human suffering and environmental degradation [damage] and must become major priorities for national and international action," said a statement endorsed by more than 100 individuals and environmental, population and family planning organizations.

Wednesday's public appeal marks the revival of a 1960s-era coalition [temporary alliance] of environmental and population control activists, many of whom later went their separate ways to pursue more specific goals. . . .

> *". . . the increase in population and in resource consumption are basic causes of human suffering and environmental damage."*

By assembling a broad group of participants, ranging from two-time Nobel Prize-winner Linus Pauling to National Organization for Women President Molly Yard, the coalition hopes to overcome the political obstacles that have blocked previous population control efforts.

"Never before has such a large and diverse group of our country's leaders endorsed one common statement expressing the vital link between population growth and the environment," said Dale Didion, vice president of the Humane Society of the United States.

The United States, whose citizens represent 5% of the world population yet consume 25% of the world's resources, and other industrial nations are just as responsible for the harmful effects of overpopulation as are rapidly growing developing nations, Didion said.

Besides contributing to poverty and hunger, overpopulation has resulted in global pollution, deforestation, resource depletion [the using up of resources] and plant and animal extinction, the coalition said.

The group called for developing better family planning services and improving education throughout the world. It noted that women with seven or more years of schooling have an average of 2.2 fewer children than uneducated women.

Representative Chester G. Atkins (D-Massachusetts), chairman of the Congressional Coalition on Population and Development, faulted U.S. leaders for taking a passive approach to overpopulation.

He cited the 1989 International Forum on Population, at which the United States "came shackled [held back] by its own misguided public policy, unwilling to support the most simple and effective solution to population growth: voluntary family planning services."

Atkins said that, in the next generation, more than 3 billion women will come of child-bearing age. Unless efforts are made to control population growth, "it will create the most monstrous, uncontrollable, devastating environmental crisis this planet has ever known."

1 GETTING INVOLVED

Decision Making

Find Out More The article lists six problems that may be caused by increasing population: world poverty, lack of food, global pollution, deforestation, resource depletion, and extinction of plants and animals. Work with a group of six students, and have each student choose one of these problems. Go to your school or local library and research to find out whether the problem you have chosen is actually caused solely by overpopulation or whether it's one of the contributing causes. When each member of the group has completed the research, discuss your findings. Be prepared to present your findings and defend your reasoning in a class discussion.

Gathering Information

Write a Letter Work with four other students to write a letter to the Congressional Coalition on Population and Development in care of the U.S. Senate. As a group, try to come to a consensus. In the letter, state your group's views on world population and what, if anything, your group thinks should be done to control population.

Cooperative Learning

Create a Public Service Announcement Working in a group of three or four students, create a 30-second public-service announcement (PSA) about world population for your school public address system. Your PSA can be recorded on audiotape. Ask your teacher to arrange for your PSA to be broadcast throughout the school. You might also want to organize a contest and have the class's public service announcements judged to see which one is most effective.

2 FOCUS ON...
A World Becoming More Crowded

Canada is a large country with a relatively small population. In the past, Canadians have not worried about overpopulation. As this article shows, however, population is a global issue. No country can afford to ignore world population growth.

BEFORE YOU READ

Work in a group of four students to find out how much you know about overpopulation before you read the article. Discuss what the word *overpopulation* means to you. What countries or areas on the list below do you think are overpopulated? What areas are not highly populated? Why do you think this is so? Now think of some U.S. cities that are highly populated. Organize your knowledge in three lists: very populated areas, not so populated areas, and highly populated U.S. cities.

Areas

China	India	Australia
Canada	Alaska	Sahara
Mexico City	Siberia	Antarctica

WHILE YOU READ

This article presents a number of facts and figures. To help you organize the information, create a chart like the one shown here and fill it in as you read.

Overpopulation: How it affects...			
environment	food	health	economics

95

Running Out of Room

by Patricia Orwen
From: *The Toronto Star*
May 26, 1991

[Two thousand years ago] there were about 1 million people on Earth. From there, women begat [produced] sons and daughters who in turn begat more sons and daughters until in 1900, . . . we were 2 billion. Then suddenly it was as if someone pressed the world's fast-forward button.

Our numbers hit 3 billion in 1960; 5 billion in 1986; today, we're at 5.4 billion. Every year, we add another 95 million people. That's more than three times the population of Canada in a 12-month period or 260,000 every 24 hours. In the time it takes to read this sentence, the world's population grew by another 30 people.

This unprecedented growth—3.4 billion people in one lifetime—is one of the most crucial events in the history of our life on this planet, says Edward Pryor, director-general of the census and demographic statistics [statistics that describe the characteristics of populations] branch of Statistics Canada.

"It's hard to even comprehend numbers like these," says Pryor. "But they have the potential to change virtually every aspect of our lives . . . our economy, the environment, everything. The very stability of the world is in question."

Canadians tend not to think about this issue, says Pryor, because it's not happening here. It's really a visible problem only in other parts of the world. We can expect that the information collected from new census data June 4 will show that Canada is growing relatively slowly, about .8 per cent a year. The rest of the world, meanwhile, is increasing by about 1.7 per cent annually—more than twice as fast as us.

"With this kind of growth, we're no longer able to look just at our own situation," says Pryor. "With each passing year, we are more affected, more drawn into what may well be a global crisis."

Pryor and others warn that the 1990s may be our last chance to halt the disastrous upward spiral of our numbers. The United Nations estimates we'll reach 6.4 billion by the end of the century, 8.5 billion by 2025.

Beyond that, estimates range widely. The Washington-based Population Crisis Committee, a non-profit educational organization, says that if the availability of birth control isn't improved dramatically in the '90s, by the end of the next century our numbers will grow to 27 billion.

"These are really outer-space numbers . . . it's practically impossible to envision them," says Hugh O'Haire of the United Nations Population Fund. "But just look around your office or your city or town and try to imagine two or three times as many people around you. Imagine what that would mean in terms of the air, the forests, even the noise. Can you imagine 14 billion people all with their boom boxes tuned to rock and roll? . . . That could be what's in store for us in the future."

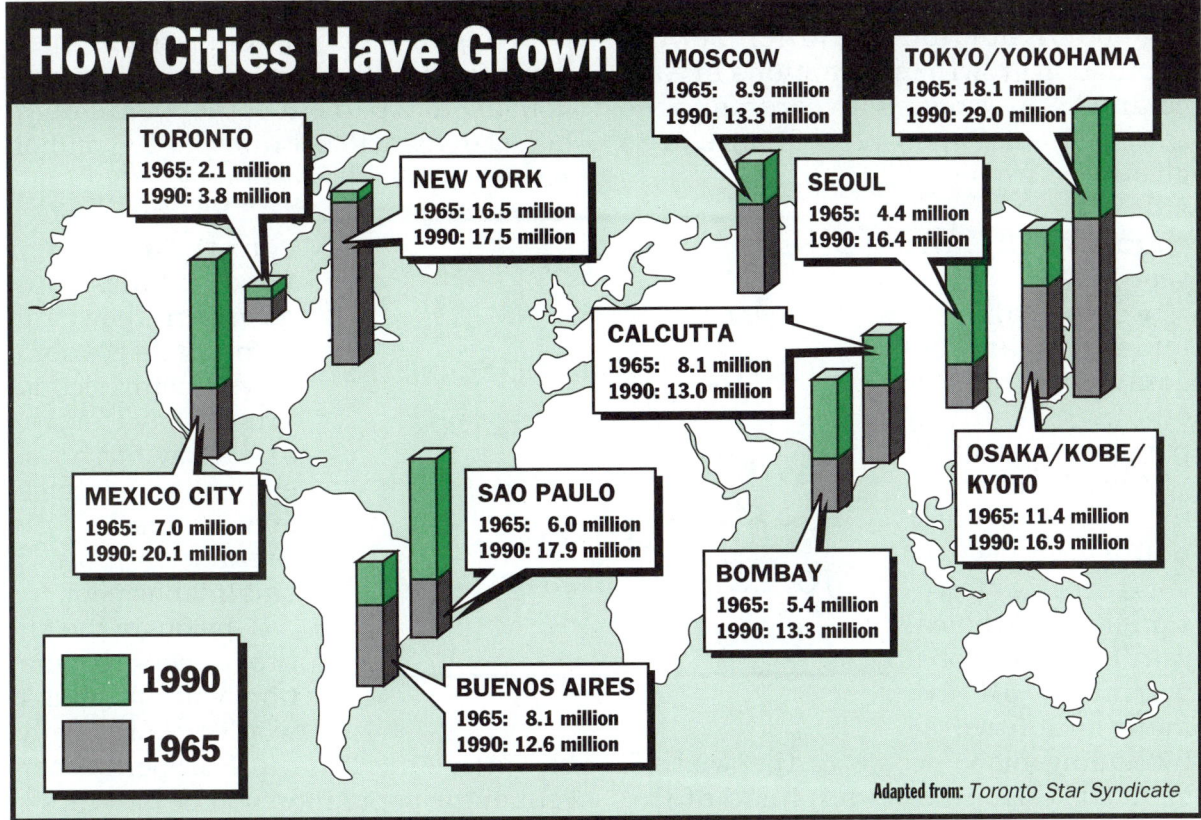

Long before then, the population explosion, or population bomb as some call it, would have devastating effects.

The more of us there are in the world, the more demand there is for such basics as air, food and water. And the faster our numbers increase, the harder it is to make adjustments to accommodate those people.

"At some point," says Sharon Camp of the Population Crisis Committee, "there must be limits to the world's carrying capacity." . . .

According to the United Nations:

*Some 580 million people—more than 20 times the population of Canada—are living in absolute poverty, which is defined as lacking the basic necessities of life. Many of these people have moved to cities, resulting in explosive urban growth in almost all parts of the developing world.

The most notorious [widely and unfavorably known] of these cities is Mexico City, whose 20 million inhabitants struggle to exist in what some consider an urban nightmare of rat-infested dumps, noise, traffic and ever-present smog . . .

*More than 85 countries have urban populations double those of 10 years ago. Rapid urbanization [growth of cities] has become the dominant demographic trend of the late 20th century. By 2010, half of the world's population may live in cities, says the Population Crisis Committee.

Apart from the escalating [increasing] need for housing, water and sanitation services, there is the urgent prob-

Populations HUMAN POPULATIONS **97**

lem of how to feed these people.

Food production has fallen behind population growth in 42 nations in Africa and Latin America. According to the U.N., 486 million people will have to be given food from outside their own countries by the year 2000.

Ecological problems, too, are becoming more serious as a result of population growth.

"Whether we're talking about water pollution, water shortages, soil erosion, loss of tropical rain forest or global warming, all are made much worse by adding more people to the world," says Dr. Joseph Speidel, head of the Population Crisis Committee. . . .

At the moment, industrial nations such as Canada are responsible for about 75 percent of the world's fossil fuel consumption. But developing countries are quickly catching up. By 2025 they are expected to produce 16 million tonnes [tons] of carbon dioxide a year, three times the present level, most of it through fuel consumption. It's this increase in carbon dioxide that is the principal cause of global warming.

Not only do we need to be concerned about how many more people we'll have in the world, we also need to look at how we are distributed by age.

By the turn of the century, the number of the world's workers aged 15 to 64 will grow by 389 million or 16.5 percent, according to the International Labor Office in Geneva. But the vast majority of those workers, 244 million, will be living in Asia. By contrast, Europe's work force will increase by only 6 million, about 2 percent of the overall number. Canada will see another 2.6 million workers.

The size of a country's labor force is as important as buying power in defining its economic well-being, economists say. Aging and a low birth rate can make a society economically dependent upon a small labor pool.

Obviously the entire world's population will age, but it won't age uniformly. In North America and Europe, 20 percent of the population will be 65 years of age and over by 2025, while in Africa the number will be less than 4 percent.

Within 10 years, 100 German workers will support 31 elderly dependents. In Canada, 100 workers will support 20 retirees. The same number of workers in Singapore will support only 13 seniors.

While low birth rates may have boosted living standards in Europe and North America in the short term, they will contribute to declining prosperity in the future unless there are very substantial gains in productivity.

Within a generation, says Pryor, the population of Africa is expected to increase by 184 percent. Combined with the growth projected for Asia, the world stands to gain another 3.5 billion people, bringing the total population to 8.5 billion.

> *"Try to imagine two or three times as many people around you. Imagine what that would mean in terms of the air, the forest, and even the noise."*

"How will financial and social support be provided for a world population in excess of 900 million aged 65 and over when the workers required to provide that support live in Third World nations . . . (and) the resources, the technology and the prosperity remain in the developed world?" asks Pryor.

Canada, he says, could be under great pressure in the future to bring in more young immigrants to bolster our diminishing labor force.

What we should be doing, however, he says, is becoming involved in global training programs, expanding our universities into developing nations so that the next population wave is able to sustain the global economy.

We must also continue to collect population information in order to be able to make decisions on not just a national scale, but a global scale.

Between now and June 4, Canada's census day, more than 60 countries, including Switzerland, Spain, China, Brazil and Syria, will be counting people. A total of 207 countries have had at least one census since 1950.

Such global statistics are fundamental for addressing many issues that are no longer unique to any one country, says Pryor.

2 GETTING INVOLVED

Critical Thinking

List the Changes The dramatic increase in population predicted in this article will create many changes in your own lifetime. Work with a group of students to predict some of these changes that may take place in your daily life by the year 2025. List one change related to each of the following human needs: food, water, housing, sanitation, recreation, health, education, future employment. What measures could be taken to deal with these changes? Present your group's findings to the class.

Decision Making

Take a Stand At the top of a piece of paper, write the following statement: "Population growth is a visible problem only in other parts of the world." Then write a brief essay in response to the statement, either agreeing or disagreeing with it. Be sure to back up your ideas with facts from the article, your own opinion, or from additional research.

Decision Making

Make a Persuasive Speech A persuasive speech is a speech that tries to influence the attitudes, beliefs, or behavior of an audience. You make a persuasive speech in order to change people's minds or move them to take some action. Imagine that you are an Ambassador of a country in which the population is increasing dramatically. Suggest a plan for a program that would reduce the population growth. Explain how your plan would be implemented. Make sure you present supporting facts and examples as evidence of your plan.

3 FOCUS ON...
Successes in Population Control

Population control is an issue on which everyone seems to have an opinion, and the author of this article is no exception. He wrote it not just to inform, but to persuade. As you read the article, see if you can identify where he is being persuasive. How successful is the writer at persuading you? Do you think overpopulation is a big problem?

BEFORE YOU READ

"Population is rapidly taking center stage as an issue of the 1990s." This sentence is found in the article you are about to read. Using words from the list below, write at least four sentences that could appear in the article in support of this viewpoint. You can use more than one of the words in each sentence. When you have finished, compare your sentences with those of a partner.

Words

environment
population policies
national
family planning
economics
health care
ecological
illiterate
government
industrialized
billion

WHILE YOU READ

The article mentions several countries where population control has been successful. Below is a chart that lists the factors of a successful program. As you read the article, find the details that explain why each factor is an important part of the program's success.

Countries Where Population Policy Works	
use of population information	
richer countries' roles	
social development	
attitudes toward women	
your ideas	

Busting The Boom: Population Control Works

But continued progress needs more funding

by Alex Marshall
FROM: *WORLDPAPER*
APRIL, 1991

Harare, Zimbabwe— . . . Only a few years ago, giving away condoms in public in Zimbabwe, or any African country, would have been unthinkable. Ten years ago, virtually no one in Zimbabwe used family planning. Now, the figure is nearing 40 percent of women between the ages of 15 and 44, according to Florence Chikara, chief of the FPC's information unit.

"Nearly all of Zimbabwe's women, 96 percent, know about family planning," says Chikara. "The problem is soon going to be, how do we get the services to everyone who wants them?"

Population is rapidly taking center stage as an issue of the 1990s. In nearly all developing countries, there is an urgency to restore demographic balance —for ecological as well as economic and social reasons.

Population information is vital to national planning; programs acting on that information—for example, to increase the use of family planning, lower fertility rates and slow urban growth—can and do succeed, under the right conditions.

All of the East and Southeast Asian nations which have been so economically successful in the last two decades started their resurgence [comeback] with strong population policies, as part of a package of social and economic reforms. Parts of south Asia which have been most successful in social development, notably the state of Kerala in India and Sri Lanka, made family planning an integral part of their social services and made it accessible [available] to women and men.

South Asia shares with Africa the highest rate of growth and the lowest economic indicators. The countries of the region have all had family planning programs for many years; most with only limited success. But Kerala and Sri Lanka show that success is possible here as well.

Parallel to the governments' understanding of the necessity for action in population, there is a growing understanding among women—and some men—even among the poor and illiterate [unable to read], that family planning offers a safe and affordable alternative to childbirth, and there is a growing demand for services. As a result, there is increasing pressure on governments to respond to increased demand by improving services, and also to help change social conditions which discourage the use of family planning.

One of the principal obstacles is the age-old belief that women have only one role in society: to bear and raise children. This prejudice against freedom of choice for women is deep-rooted; but it is based on a myth and always has been. Women in all societies have been active as food producers. They have been manufacturers of woven goods, pots, baskets and all kinds of other articles and they have traded in the results of their work. As well as sustaining the family and national economy, they have provided many of the essential services from education to health care. From their close connection with forest and field and the products of earth and water, they have effectively been managers of their microenvironment [small, local habitat].

> **"Industrialized nations contain one-quarter of the world's population."**

These contributions have extended far beyond the primary or traditional role as mother and housekeeper; yet they have never been recognized either traditionally or in modern development planning, with the result that women have become increasingly marginalized [scarcely considered] even in rural societies. In the altered circumstances of urban life, women have been able to assert themselves. But even here they are faced with the results of ancient prejudice.

It is absolutely essential that this be changed. One of the keys to change is the achievement of reproductive freedom—the freedom from which all other freedoms flow. Success in this is a condition of successful sustainable [when resources are being renewed quicker than people are using them up] development programs, including population programs and environmental security.

These issues will become more urgent as we reach the 21st century. In absolute numbers, population growth is faster now than ever before, and the fastest growth is in the poorest countries. The recent conference on the least developed countries identified particular population needs, especially with respect to environmental degradation. In a typical developing country, nearly half the population is under 15. They are all potential contributors to development, but they need food, housing, health care, education and productive work before they can make their contribution. The demand in developing countries for all these services is vast and will continue to rise. There is an urgent need for action now, not only to meet present needs but to anticipate the needs of the future.

This action must include greater attention to slower, more balanced population growth. We are now approaching a global population of 5.4 billion, and it is increasing at 92 million a year. Merely in order to maintain progress towards the United Nations' median projection of 8.5 billion people in 2025, we need to ensure that the number of family planning users goes up by two-thirds in the next decade, from 381 million couples now to 567 million in the year 2000.

Otherwise there is a possibility that we could reach 10 billion by 2025 and 14 billion eventually.

Such a huge increase in the use of family planning is not out of the question; indeed it is a fairly modest target. In money terms it means doubling the amount spent on population programs to $9 billion annually—but that is only 1 percent of the global arms bill, to make one obvious comparison.

It will, however, need both understanding and commitment: understanding that it is not a choice for the international community, but a necessity, and commitment from all levels of society, particularly from the governments of developing countries.

The resources of developing countries are limited and their choices hard to make. They already find the bulk of the funding for family planning programs from their own resources. They will be greatly helped if the richer countries will increase the assistance level. In cash terms the estimated increase is fairly modest—from $675 million today to $4.5 billion by the end of the decade —and well within the capabilities of the industrialized countries.

The richer countries have a dual role in helping to balance environment, resources and population. With only one quarter of the world population, they are responsible for about three-quarters of the energy used and waste generated; they must not only hold down their own demands on world resources but be willing to transfer to developing countries the technology which will make "clean" development possible. Demanding that developing countries limit population growth demands equally responsible behavior from industrialized countries.

The 1992 World Conference on Environment and Development (WCED) will be most important in establishing international priorities into the next century. Population was not on the agenda in the first conference in 1972, but we hope that the issue is now one of consensus rather than controversy, and that it will get its due measure of attention as one of the central items on the agenda. Eventually, all such conferences boil down to a discussion about the human future. What could be more relevant to the subject than how many we shall be, where we shall live and whether our rate of growth and concentration is in a sustainable balance with our

"Today's problems should have been solved in the 1950s, but in the 50s we were solving the problems of the 20s, in the 20s we were solving the problems of the 1890s . . ."

physical environment?

The 1992 conference will be followed by a world population conference in 1994, at which the international community will assess progress and assign priorities for the next 10 years. Solid input from the WCED will greatly improve the quality of the discussion at the 1994 conference, and its practical value to developing countries.

3 GETTING INVOLVED

Gathering Information

Write a Letter to Request Information A letter to request information is a letter you send to a person or organization, asking for specific information. As a class, write to one or two of the following organizations: *Zero Population Growth, Population Institute*, and *Population Reference Bureau* and request information on population growth. (The addresses can be found in the Community Resource Directory starting on page 244.) Make a bulletin board display with the information you receive.

Cooperative Learning

Debate the Issue Work with four other students to debate the issue of family size restrictions and government sponsored family planning. One person should be the moderator, two students should take the stand for restricting family sizes, and two students should take the stand opposing the restrictions. Using your own opinion prepare a one-minute statement concerning this issue. Develop at least three questions to ask the opposing group. After each opening statement has been given, continue to debate the issue. After ten minutes, the moderator should summarize what was said and decide which side presented the most convincing argument.

Cooperative Learning

Create a Concept Map Work with another student to create a concept map to show what might happen if the world's population growth does not slow down. First, brainstorm to create a list of primary and secondary effects. When creating your list, be sure to include ecological, economic, and social effects. Also indicate what specific groups of people will be affected by the dramatic increase in population. Your map may be similar to the one shown. Draw lines between the concepts that relate to one another. Compare your concept map with the map drawn by another pair of students.

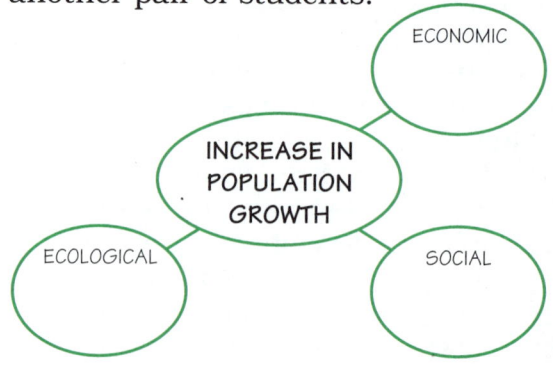

4 FOCUS ON...
Birth Control and Norplant

Seldom has an issue become controversial as fast as Norplant did. Before it was even on the market, a judge had ordered that a woman convicted of child abuse use it to prevent her from having further children. What is it about Norplant that makes it so controversial? How is it different from other methods of birth control? These two articles raise at least as many questions as they answer.

BEFORE YOU READ

The chart below lists some words used in these articles that may be unfamiliar to you. Copy the chart onto a separate piece of paper. If you can define the word, place a check mark in that column. If you have seen or heard the word but aren't sure what it means, place a check mark in that column. If you are completely unfamiliar with the word, place a check mark in the ? column. In the last column write the definition of the word if you know it, or look it up in a dictionary.

Word	Can define	Have seen/heard	?	Definition
probation				
debate				
implant				
plea agreement				
reproduction				
contraception				

WHILE YOU READ

The articles focus on two different ways Norplant can be used: as forced contraception and as a method of birth control. Copy a diagram like the one below on a separate piece of paper. As you read, list the advantages and the disadvantages of each way of using Norplant.

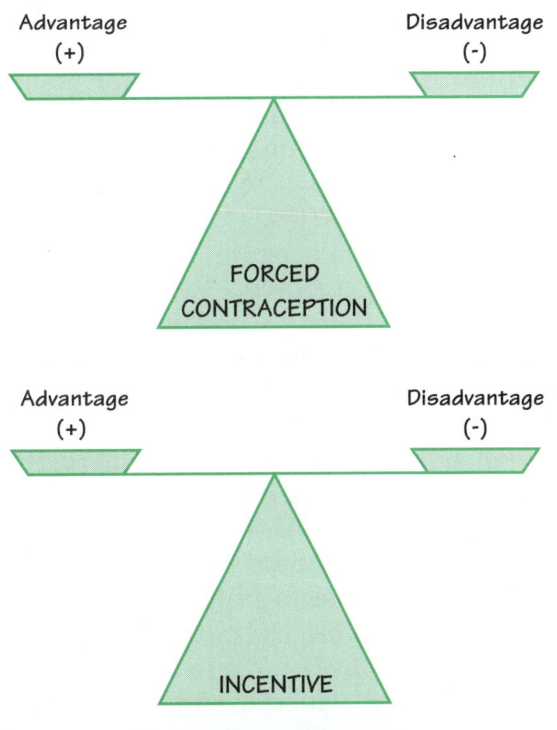

105

Norplant Renews Debate Over Forced Contraception Reproductive Rights

by Tamar Lewin
FROM: *THE MORNING CALL*
JANUARY 13, 1991

Less than a month after the federal government approved a new birth control device that is implanted [placed] under a woman's skin, the long-lasting device is the focus of a renewed debate over forced contraception.

A county judge in California has already ordered that a woman convicted of child abuse use the device for three years as a condition of probation. Experts in medical ethics say that because of the ease in using the device, which is not yet on the market, other judges may be tempted to order its use in cases where women are seen as unfit to be parents.

The device, Norplant, was approved by the Food and Drug Administration on December 10 [1991] and was widely hailed as a "dream method" of birth control because it can be easily implanted in a woman's arm, and then remain effective for up to five years.

The first substantially new contraceptive in 25 years, Norplant consists of several soft, matchstick-size rubber tubes that are placed under the skin of the woman's upper arm, where they release the female hormone progesterone, one of the components [ingredients] of birth control pills.

With the exception of sterilization, Norplant is expected to be the most effective contraceptive, because it does not depend on a person's remembering to use it.

"Norplant presents a special temptation to judges because it's so long-lasting and doesn't require any cooperation after it's implanted, and can be monitored [checked] by a parole officer just by looking at the woman's arm," said Dr. George Annas, director of the program on law, medicine and ethics at the Boston University School of Medicine.

"I think we're going to see more of these cases. It's kind of amazing that this has happened already, when hardly any physicians even know how to implant this thing."

In the California case, Tulare County Superior Court Judge Howard Broadman last week ordered the implantation of the device in Darlene Johnson, 27, a pregnant mother of four who pleaded guilty to beating two of her children with a belt.

The order was issued at her sentencing, with no notice to either the woman

NORPLANT® IMPLANTS

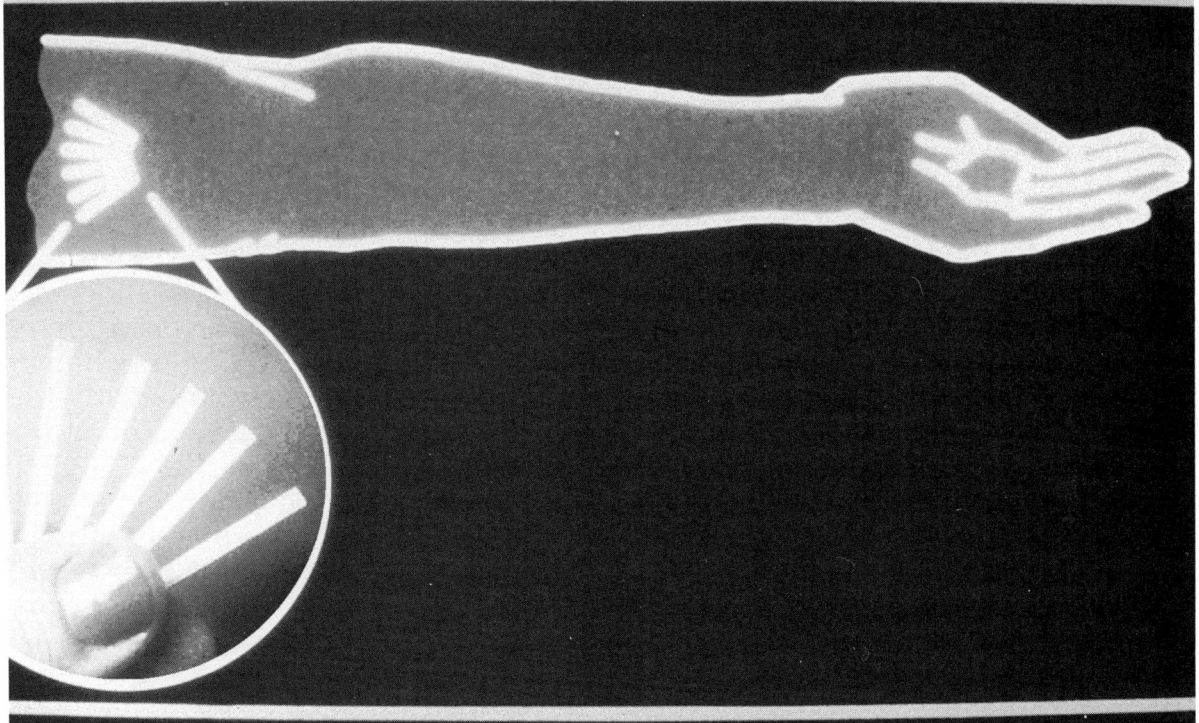

These tiny tubes of Norplant are inserted just beneath the skin in a woman's upper arm. As soon as Norplant is in place, it begins to release a synthetic hormone, progestin, into the bloodstream. The hormone prevents pregnancy for approximately five years.

or her lawyer, Charles Rothbaum. Broadman reconsidered the order at a hearing Thursday morning on a motion filed by Rothbaum, but refused to rescind [take back] his order.

Rothbaum said his client had been completely taken by surprise by the judge's sentencing order. In a plea agreement arranged earlier, Johnson was to be sentenced to one year in jail and three years of probation.

He said she had agreed to the judge's order only because she was afraid that if she refused, she would go to jail for four years....

When Norplant was first introduced last month, medical ethicists warned that the device was so attractive that it might be abused by those seeking to force certain groups of women, particularly retarded women or women receiving welfare benefits, to forgo having children.

A Dec. 12 editorial in *The Philadelphia Inquirer,* with the headline "Poverty and Norplant—Can Contraception Reduce the Underclass?" suggested that because of growing poverty among blacks, welfare mothers should be offered incentives to use Norplant.

But so many members of the *Inquirer's* news department denounced the editorial as a racist endorsement of genocide [destruction of an entire group] that the paper took the unusual step of printing an apology.

In the California case, too, Johnson's status as a welfare recipient may have played a role.

While those who support such sentences are not organized, the number of sentences seems to be growing.

"There's definitely a trend toward third-party involvement in reproductive decisions, including all the attempts to put women in jail for taking drugs that can affect the fetus," said Arthur Caplan, the director of the University of Minnesota's Center for Biomedical Ethics.

"And I know of at least three cases, none of which got very far legally, where women were threatened with compulsory abortion, either because of a history of child abuse, or because they were taking drugs that could harm the fetus."

The embryologist [scientist who studies the formation and development of embryos] who developed Norplant, Sheldon Segal of the Rockefeller Foundation, said he was distressed by the Johnson case and other suggestions that the device might be forced upon some women.

"I just don't believe in restricting human rights, especially reproductive rights," he said. "And I'm also bothered because this is a prescription drug, with certain side effects and certain groups of women for whom it may not be appropriate."

"How does the judge know if the woman is diabetic, or has some other contraindication [bad reaction] to the drug? That's not his business."

Rothbaum said Johnson, who is now seven months pregnant, was diabetic and thus not a good candidate for Norplant.

In recent years, there has been increasing interest in chemical castration of rapists, despite a 1942 ruling by the U.S. Supreme Court that struck down an Oklahoma law permitting castration for repeated felonies involving "moral turpitude [corruption]."

In a few scattered cases across the country, judges have offered men convicted of rape or sexual abuse a choice of chemical or surgical castration or long prison sentences. Such choices have generally been overruled or withdrawn before castration actually occurred.

But in 1988, Melody Baldwin, 30, an Indiana woman with a history of personality disorders, agreed to be sterilized as part of an agreement and pleaded guilty to killing her 4-year-old son with an overdose of psychiatric drugs prescribed for her.

"This kind of thing happens a lot in lower courts and never gets challenged because the defendant's happy not to be in jail," said Annas.

"A lot of people have given up on social policy, on taking care of poor women, and there is an increasing undercurrent that since we don't really know what to do about crack addicts, people with AIDS and child abusers, we should stop them from having kids." . . .

"There is some latitude for creative sentencing," said Rachael Pine, of the American Civil Liberties Union's Reproductive Freedom Project. "But where you're talking about someone having surgery to be sterilized or implant a contraceptive, you've clearly crossed the line." . . .

Birth Control or Woman Control?

Norplant not meant for coercion

by Ellen Goodman
FROM: *CHARLOTTE OBSERVER*
FEBRUARY 19, 1991

Dr. Sheldon Segal expected Norplant to generate a controversy sooner or later. It was the "sooner" that took him by surprise.

On the very morning the FDA approved the long-lasting contraceptive implant, Segal found himself in a taxi between television studios listening to someone on a radio talk show loudly proclaim that every girl should have it stuck in her arm at puberty. The cab driver uttered his full-throated agreement and the man who developed this new birth-control method shrank down into his seat: "That was Day One," Segal says.

On Day Two, *The Philadelphia Inquirer* published an editorial about Norplant saying that readers should "think about" Norplant as a tool in the fight against black poverty. The message, spiked with a volatile mix of race, class and contraception, kicked up a storm.

Segal sent off his own outraged letter to the editor. But before it was published, the story struck again. A California judge ordered a convicted child-abuser to use Norplant as part of her sentence.

The contraceptive wasn't even on the market yet.

Sitting in his office at the Rockefeller Foundation, he shakes his head at all this. "We created a method to enhance reproductive freedom and people keep finding ways to use it for the opposite purpose."

It took 24 years to develop, test and approve an implantable device that can prevent pregnancy for as long as five years. It took less than two weeks for Norplant to be billed as a new method of coercion.

Yes, the team that worked on Norplant had been concerned that a government would misuse the device to enforce birth control. But frankly, they were worrying about China, not California.

Now the story has moved to Kansas. Last week, the legislature held hearings on a bill that would pay welfare mothers $500 to get the implant. It would also pay for the Norplant, plus an annual checkup and a $50 check a year.

Under the bill, the state would offer an incentive to one class of women—poor, single mothers on welfare—for one kind of birth control: Norplant.

The man who came up with this idea, Kerry Patrick, a Kansas state representative, describes himself as "a pro-life Republican Presbyterian." He defends this bill as a way "to encourage people to engage in a certain type of behavior."

At the same time, he figures to save the state the $205,000 it costs for each child on welfare from birth to adulthood. . . .

Other governments attempt to influence the decisions families make about fertility all the time. In France, they give bonuses for each baby, In India, they offer "expense money" to citizens who get sterilized.

The Kansas offer of $500 plus free birth control may sound like a good deal for a poor woman who wants Norplant. But "the line between incentive and coercion gets very fuzzy," says Segal. The $500 bonus can be a heavy government hand on the scales of choice for the poor. He worries that "when you single out a welfare mother, wave a $500 bill in front of her face and say that the government is going to induce you not to have children, you've gotten into a risky area, ethically and morally." . . .

4 GETTING INVOLVED

Cooperative Learning

Debate the Issue   Organize a debate on this question: Should judges or juries be allowed to order women to use Norplant as part of their sentencing? For the debate, you will need two teams and a moderator. Each team should prepare an opening statement and be ready to answer questions. The moderator asks the questions and allows both teams equal time. (Students who are not on one of the teams may help the moderator prepare the list of questions to be asked.) When the teams have answered all the questions, they end with a concluding statement. The entire class then votes to see which side was most persuasive.

Decision Making

Write a Position Statement   A position statement recommends a particular decision or course of action and the reasons why the action should be taken. Write a position statement in favor of financial incentive programs for women to use Norplant. Then write a second position statement, taking the opposite point of view. Be sure to back up both positions with logical reasoning.

Gathering Information

Develop a Public Opinion Survey   A public opinion survey is a tool used to gather information about people's attitudes. You take a survey to make generalizations about the attitudes of a group of people. Work with three other students to develop a public opinion survey to determine people's opinions on the appropriate uses of Norplant. You will need to word the questions carefully to ensure a fair response. Have each person in the group ask five people to complete the survey. As a group, analyze your results and organize your data in a table. See page 236 for some helpful hints on organizing data in data tables.

5 FOCUS ON...
Abortion and People With Disabilities

Is knowing that a child will be severely disabled an appropriate reason for a woman to choose an abortion? Anti-abortion groups are uniting with advocates for the disabled to say No. This article points out many of the difficulties and contradictions involved in taking a position on this question.

BEFORE YOU READ

Many people in our society with disabilities are able to conduct a relatively normal lifestyle. There are groups that help these people find jobs, live independently, play sports, and so on. The diagram names some things of interest to people with disabilities. Why would each of these be of great concern to them?

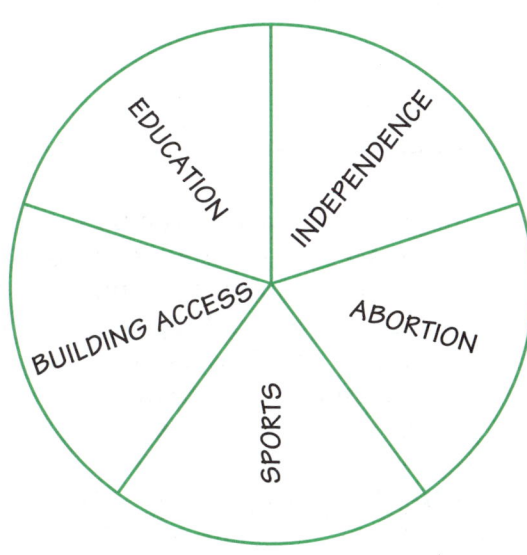

WHILE YOU READ

The article discusses some concerns of anti-abortion groups and groups for persons with disabilities. As you read, list the views shared by both groups on a chart like the one here. It is conceivable that these groups will form an alliance in the near future. In addition, list some of the possible political consequences of such an alliance.

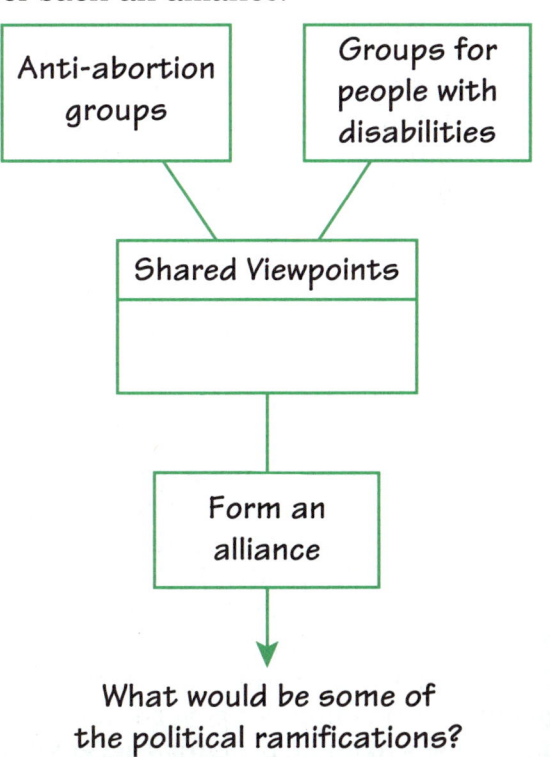

Abortion Issue Divides Advocates for Disabled

by Steven A. Holmes
From: *The New York Times*
July 4, 1991

Washington—Barbara Faye Waxman would cringe when she heard her co-workers at a Planned Parenthood clinic in Los Angeles discuss prenatal screening and the need to abort a disabled fetus. For Ms. Waxman, who must use a wheelchair and needs a respirator to breathe because of a neuromuscular impairment [problem with the nervous and muscular systems], such conversations were painful. . . .

At a time when their political strength is increasing, advocates for the disabled find themselves torn by conflicting emotions over the issue of abortion. Anti-abortion groups, sensing that ambivalence, are courting advocates for the disabled more and more.

Should such a union occur, it could create a force that would have a significant effect on the politics of abortion.

What They Have in Common

In June, for instance, the National Right to Life Committee, the nation's largest anti-abortion group, elected Robert Powell, a quadriplegic from Galveston, Texas, as its vice president. Mr. Powell is a co-founder of the Galveston Coalition for Barrier-Free Living, a group that has pressed government and businesses to make buildings accessible to disabled people.

In reaching out to organizations representing the disabled, anti-abortion leaders hope to build on the two groups' common opposition to euthanasia [mercy killing].

"People are saying that the disabled should have a certain quality of life, and short of that we are just a drain on scarce resources," said Lillibeth Navarro, the southern California orga-

> **"** *We had a task force study the issue, but we couldn't take a position.* **"**

nizer for Adapt, a Denver-based group that has strongly pressed for greater rights for individuals with physical impairments. "The thinking that goes into abortion and euthanasia are very much the same," added Ms. Navarro, who also is a member of Feminists for Life, an anti-abortion group.

In the last two years abortion opponents have succeeded in enacting laws in Louisiana, Pennsylvania, Utah and Guam that would severely limit abortions if appeals do not succeed. Likewise, the disabled have been flexing their political muscle, gaining passage of the Americans With Disabilities Act last year. The law bars discrimination against people with disabilities, and will eventually affect virtually every commercial establishment in the country.

Trying to Forge an Alliance

"The anti-choice movement gravitating toward the disabled is just an attempt to forge an alliance that will strengthen them," said Kate Michelman, executive director of the National Abortion Rights Action League. "They need to increase their emotional support among people. Any time they can play on the emotions, they feel they can gain some support."

Some who lobby for the rights of the disabled scoff at the idea of a coalition with anti-abortion groups and express skepticism about the commitment of anti-abortion groups toward the plight of the disabled. They say no anti-abortion group lobbied for legislation barring discrimination against people with disabilities or to provide support for the parents of disabled children—steps they say would reduce the pressure on a pregnant woman to abort a fetus found to be impaired.

"If they truly cared about disability, where were they on the Americans With Disabilities Act or the Fair Housing Act or the Civil Rights Restoration Act?" asked Pat Wright, head of government affairs for the Disability Rights Education and Defense Fund, a group based in Berkeley, California, that litigates and lobbies on behalf of disabled people. "They were nowhere to be seen, nowhere to be heard."

Still, many advocates for the disabled acknowledge that abortion is a wrenching emotional issue.

Having fought for the civil rights of people with physical and mental disabilities, many leaders of disabled groups say they are uncomfortable limiting the rights of anyone, including those of a woman to end a pregnancy. On the other hand, they have a visceral sense [gut feeling] that if that same right to an abortion had been widely available years ago, they or the disabled children they have loved, raised and fought for might never have been born. . . .

People who lobby for the disabled are grappling with the issue when advances in prenatal diagnostic techniques reduce the time it takes to detect fetal abnormalities, thus making a woman's decision to end a pregnancy financially and emotionally easier.

Leaders of disabled groups are also aware that, along with rape, incest and danger to the life of the pregnant woman, preventing the birth of a disabled baby remains one of the reasons for an abortion that the public appears most ready to accept.

Some Avoid the Subject

Disabled people on both sides of the abortion debate agree that pregnant women who learn that they are carrying a disabled fetus should receive counseling, including information on the range of support services for children with disabilities.

Concerned that the issue could split their movement, some people who lobby for the rights of the disabled have simply avoided the subject of abortion. In recent years at least two organizations, the Spina Bifida Association and the Association for Retarded Citizens, considered adopting a position on abortion but found the issue too divisive.

"We had a task force study the issue and produce a paper, but we couldn't take the next step and take a position," said Paul Marchand, director of governmental affairs for the Association of Retarded Citizens. "And when we didn't take a stand, the collective sigh of relief was incredible."

Still, some groups representing the disabled feel that at some point they will have to take a stand. Lorelee Stewart, a vice president for the National Council on Independent Living, said her group's members would soon be polled to discern [find out] their views on the subject.

"We have been waiting for the disability community to be dragged into this debate for a long time," Ms. Stewart said. It's going to make us have to look at our politics very closely."

5 GETTING INVOLVED

Gathering Information

Find Out More The article mentions prenatal screening but does not describe how a woman might find out she is carrying a disabled fetus. Do research to find out what kinds of prenatal testing are available and what kinds of information are obtained from the tests. Are there any risks associated with prenatal testing? Present your findings in the form of an oral report.

Gathering Information

Investigate Locally Currently some states are involved in modifying their abortion laws. The U.S. Supreme Court has become involved in reviewing some of the newly modified laws. As a class, write or call your state governor's office or local representative and find out your state's laws regarding abortion, as well as the current position of the U.S. Supreme Court on this issue.

After the class has received the information, write a belief statement that summarizes your own personal feelings on the issue of abortion. Do you agree or disagree with your state's laws? You may wish to keep your belief statement to yourself, or you may wish to write to your governor stating your belief.

Gathering Information

Library Research Some potentially harmful conditions existing in a fetus can be detected by prenatal testing, however, others such as fetal alcohol syndrome and fetal drug addiction cannot. Research one of these conditions to find out the answers to the following questions. How do fetuses develop the condition? What are the effects on the fetus? What can be done to treat the child once it is born? How costly is this? If such a technology were available to detect these conditions in a fetus, some women might consider aborting such a fetus. It is also possible that the ability to detect such conditions in a fetus could lead to development of some form of prenatal treatment. Do you think such a technology should be developed? Why or why not? In giving your answer list any pros and cons you have considered.

6 FOCUS ON...
U.S. Immigration Policy

Does the United States need an immigration policy? If you've thought about this topic at all, you may have thought we have an immigration policy. According to this article, we have immigration laws but no stated purposes or goals.

BEFORE YOU READ

Below are three "word satellites" and four terms. Read the words in the top satellite and try to decide how the terms are related to each other. Discuss your ideas with a partner, if you wish. Then select the term under the satellites that best fits in with the other words in the satellite. Continue on to the other satellites and do the same thing with them. When you are finished with all three satellites, predict what the article may be about. Why did you come to this conclusion?

WHILE YOU READ

The article describes opinions about current immigration laws. Draw a framework like the one shown here on a separate piece of paper. As you read the article, write on the lines sentences from the article that support each opinion.

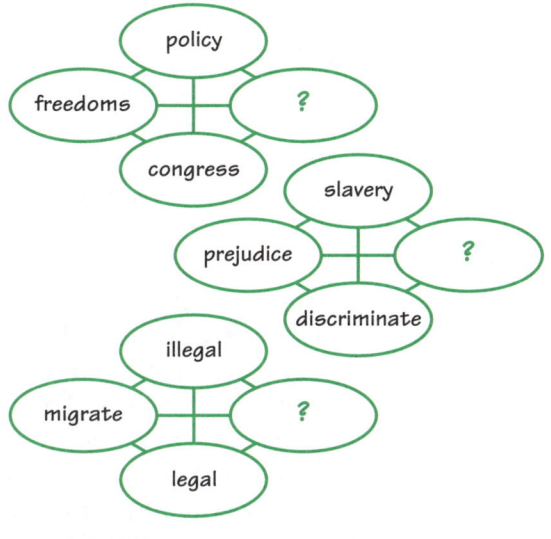

Racism   Statue of Liberty
Constitution   Immigration

> Family reunification takes priority over providing workers with needed skills.
>
> _____
> _____
> _____

> The current system can compensate for past discrimination.
>
> _____
> _____
> _____

> Large numbers of immigrants hurt native-born Americans, especially African Americans.
>
> _____
> _____
> _____

U.S. Laws Under Attack

A call for an immigration policy

by Ramon G. McLeod
FROM: THE SAN FRANCISCO CHRONICLE
JULY 4, 1991

Near-record numbers of immigrants are rapidly changing California and the nation, but critics and supporters agree that U.S. immigration laws have no clearly stated purposes or goals and may lead to unexpected results.

Current statutes [laws] are largely silent about even the most basic immigration effects, including what kinds of workers the nation needs and how many people it wants.

"If you don't know where you're going, you can end up with unintended consequences, like the very rapid population growth we're getting now from immigration," says David Simcox, director of the Center for Immigration Studies, an independent research organization.

Many analysts say Congress and the administration have deliberately avoided putting formal immigration goals in writing, fearing that any stand on the debate will inevitably draw charges of racism.

Dan Gonzales, an immigration expert at San Francisco State University, likens the lack of a formal policy discussion to the debate over slavery when the Constitution was framed. "The subject is so volatile no politician wants to have any formal position down in writing," he said.

Already, this lack of policy has led to enormous changes—and few are by design.

In the past decade alone, the arrival of at least 7 million legal immigrants and more than 2 million illegal immigrants has altered major industries, transformed [changed] hundreds of communities and shifted the political balance in major U.S. cities. And a recent change in the law will permit even more newcomers in the 1990s, most of whom will come from Asia and Latin American nations if the patterns of the past decade continue.

Immigration laws now give priority to prospective immigrants who have a family member already in the United States. So instead of the government deciding what kind of people cross the borders, critics say, the immigrants themselves decide who gets in.

At least two-thirds of the legal immigrants in the 1980s entered under "family reunification" provisions. Less than 10 percent of the visas issued in the 1980s were specifically for people who had skills needed in the U.S. labor market.

A new law this fall will boost "skills preferences" to 20 percent of all visas, but family reunification will still account for about two-thirds of entrants.

Even supporters of the law believe that the nation needs more clearly defined immigration policies and goals.

"Congress has come up with various immigration policies over time, and you

LEGAL IMMIGRATION TO THE UNITED STATES

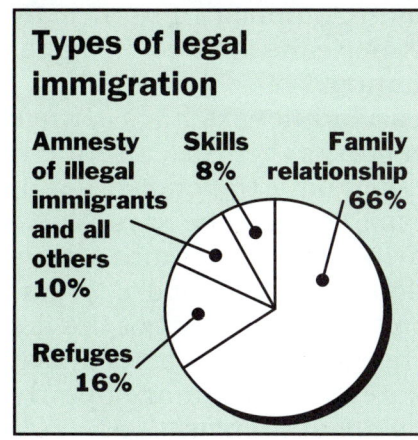
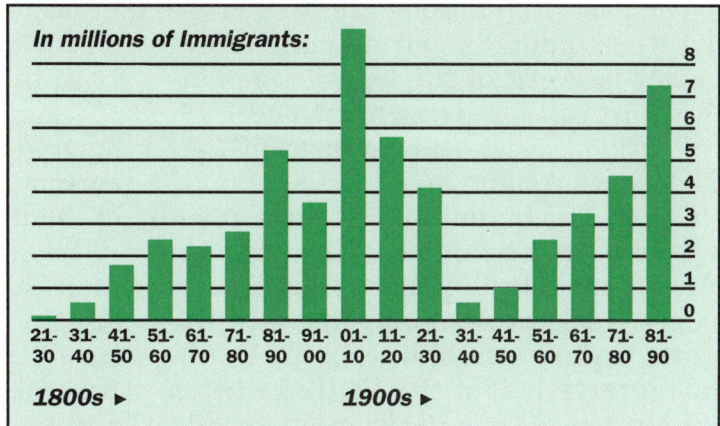

Source: U.S. Immigration and Naturalization Service   Adapted from: *San Francisco Chronicle*

can sort of guess what their intentions have been, but these usually have little connection to reality," says Bill Hing, an immigration law expert at Stanford University.

"The truth is that there has been a very poor job done predicting the impact [effect] of immigration both on the sending and on the receiving countries," Hing says.

In 1964, the late Robert Kennedy told a congressional hearing that an immigration law ending quota systems would result in at most about 5,000 people immigrating from Asia. Almost 4 million Asians immigrated after the law went into effect in 1965. . . .

Conservatives [people opposed to change] who support the current relatively open immigration system believe that more immigrants help economic progress. They find themselves in a coalition [agreement] with liberals who believe that U.S. freedoms should be open to all people in the tradition represented by the Statue of Liberty.

The opposition to current law includes liberals [people open to change] who argue that the large numbers of immigrants hurt native-born Americans, especially African Americans, and conservatives who fear that open borders will significantly alter the racial and ethnic makeup of the nation.

"You can't discuss this without someone calling you a racist," says Leon Bouvier, a liberal demographer at Tulane University. "But the fact is that important demographic shifts are going on, and to say they don't mean anything, or imply it by silence, is ridiculous."

In the absence of clear purposes for immigration, the volatile [heated] debate over how porous [open] the borders should be centers on the effects of the law as it now stands.

"It proves that the dream is still alive in the U.S.—that anyone, of any class and any race, can come in and make it if they work hard," says Ignatius Bau, chairman of San Francisco's Coalition for Immigrant and Refugee Rights.

Other supporters of the current system want the end result to be a redress of [compensate for] past discriminatory immigration law that kept out Asians and Latinos until 1965.

The past prohibitions are the reason that "persons of Asian descent represent 50 percent of the world's population, but only 3 percent of the U.S. population," says Bill Tamayo of San Francisco's Asian Law Caucus.

Opponents of the current law—most of whom argue for less immigration rather than closing the borders completely—say the system is changing the demographic makeup [population characteristics] of the United States without benefit of a decision on exactly how it should be changed.

Frank Morris, dean of graduate studies at Morgan State University, believes that the effect of immigration law is to "dilute the ratio of blacks to whites. . . . Anyone who doesn't believe that the impacts of immigration fall most heavily on African Americans ignores the history of this country."

"American employers have always favored immigrants over African Americans when they have the choice—that is the inescapable fact," he says.

Immigration backer Tamayo says many emotional statements are "a jingoistic [strong, patriotic] part of the debate," but he nonetheless agrees that Americans need to talk more openly and honestly about immigration.

"The atmosphere around this subject is so charged," he says. "I really wish people would stop yelling at each other."

6 GETTING INVOLVED

Gathering Information

Send for Information Write to the United States Immigration and Naturalization Service to find out what are the current procedures for immigration to this country. How many immigrants are allowed each year? What are the requirements for citizenship? How long must immigrants live in the United States before applying for citizenship? You also will be able to find the same information in your local library.

Gathering Information

Talk to an Expert Invite a recent immigrant to talk to your class about what the immigration process was like. Was the person on a waiting list before coming here? Did he or she have a job waiting in this country? Were there any unexpected difficulties? If the person had to learn English, how much of a stumbling block was that? How is this person's life different from what it would have been had he or she not immigrated?

Critical Thinking

Benefits and Costs With a group of three students, discuss the following question: Should immigration be tied to needs identified in the U.S. job market? To help you answer the question, list the benefits and costs of this kind of immigration law. Summarize your discussion in a brief written report.

7 FOCUS ON...
Selling U.S. Citizenship

Not everyone is happy about a new immigration program that favors wealthy foreigners. To some people, however, it's a program designed to increase investment and jobs in the United States. Which is it? Read the article and then see what you think.

BEFORE YOU READ

Below is an incomplete word map about immigration. Copy the map onto a piece of paper. With a partner, brainstorm about immigration and complete your word map.

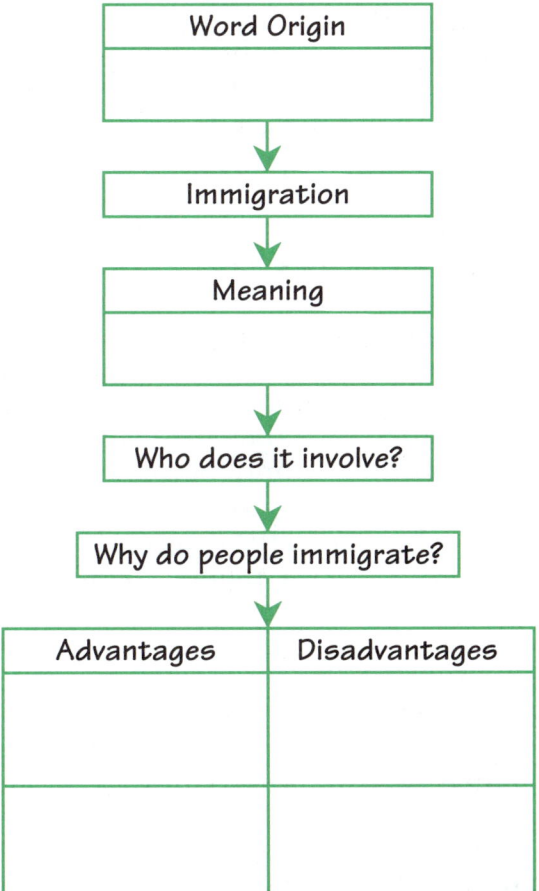

WHILE YOU READ

As you read, look for answers to these questions. Who might the "bought" immigrants be? What arguments support the program? What have been the results of similar programs in other countries? What are some problems with the U.S. program? Copy the diagram below and fill in the spaces as you read.

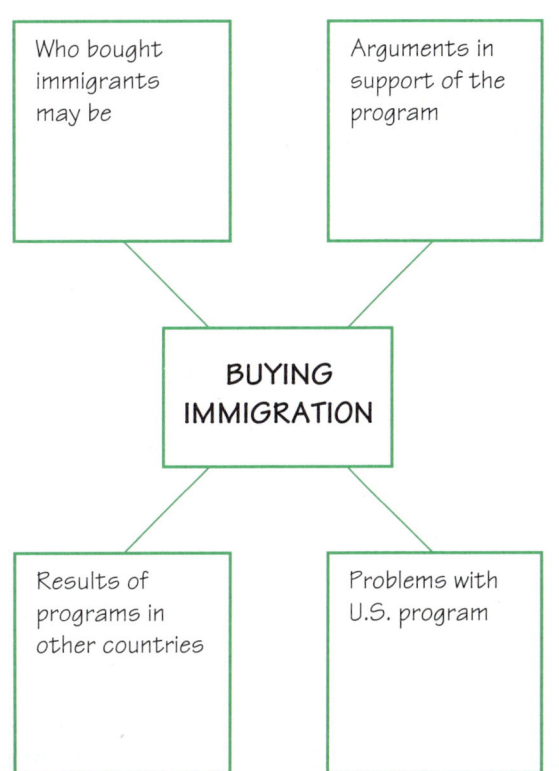

119

An Investment in American Citizenship

Immigration program invites millionaires to buy their way in

By Al Kamen
FROM: *THE WASHINGTON POST*
SEPTEMBER 29, 1991

Lady Liberty may still beckon to the "huddled masses yearning to breathe free," but Uncle Sam now extends a special welcome to those who can pay cash.

Under a new immigration program that goes into effect Tuesday, 10,000 visas for permanent residence in the United States will be available each year for those who agree to invest at least $1 million in businesses here. Full citizenship will be available for them and their families after five years.

The program, part of the Immigration Act of 1990, is an unabashed [undisguised] attempt to attract wealthy foreigners—especially Asians and, most particularly, Chinese from Hong Kong who are worried about living under communist rule in 1997.

The response—100 applications so far—has been less than overwhelming. Officials at the Immigration and Naturalization Service [INS] say they are confident many more people will apply once final regulations are issued. But immigration lawyers and others are not sure, saying the program is too restrictive and expensive.

The plan is patterned after highly successful programs in other countries, especially Canada and Australia. Canada's program, which began in 1986, has brought in more than $3 billion a year and has created more than 40,000 jobs, Canadian officials said.

Australia's program, begun in 1982, brought in $1.3 billion in new investment last year, with about 10,000 settlers coming mostly from Asia, according to the Australian Embassy.

Those figures—and a sluggish U.S. economy—overcame the discomfort some lawmakers felt about the notion of a dollars-for-visas program.

Senator Dale Bumpers (D-Arkansas), who led the opposition, thundered on the Senate floor against "auctioning off our souls" by "allowing somebody into this country simply because he or she happens to have $1 million, either inherited, made in the drug cartel, regardless of where the money comes from."

But Republican backers, such as Senator Phil Gramm (Texas), countered by quoting Calvin Coolidge that "the business of America is business" and the country needed entrepreneurs as much as any other category of immigrant. Democrats, such as Senator Edward M. Kennedy (Massachusetts), insisted the program was not selling visas, but creating jobs because, in addition to the money, investors also would have to create 10 jobs to qualify for permanent residency.

The bill's supporters predicted that about 4,000 millionaire investors, along with family members, would sign up, bringing in $4 billion in new investment and creating 40,000 jobs.

Immigration lawyers and business promoters went into a feeding frenzy at the possibilities, shuttling back and forth to the Orient, conducting dozens of seminars and discussions with prospective clients.

But now, after all the congressional angst [apprehension] and the promotional activity, immigration lawyers and business brokers are not sure wealthy foreigners will bite.

One problem, said Harold Ezell, a former INS official and now an immigration consultant in California who is advising foreign investors, is that "INS dropped the ball," and has yet to publish regulations to implement the law. As a result, applications cannot be processed. Ezell is confident applications will increase. The Canadian offer is attractive, he said, "if you want to go to Canada and freeze your buns off."

Even when final regulations are issued, possibly in the next few weeks, there still may be uncertainties, immigration lawyers said. Immigration officials are debating whether to lower the ante [required investment] to $500,000 for investments in rural or high-unemployment areas.

Entrepreneurs will receive a two-year provisional visa and, if still-to-be-decided criteria are met, a permanent green card may be issued, with citizenship three years after that. One question unanswered, however, is what happens to the visa if after two years the business falls on hard times and employs only seven people or if other regulations aren't met.

"One million dollars is not chump change," said St. Louis immigration lawyer George Newman. "People with that kind of money didn't get it because they are idiots," he said, and they are not going to jump into the program without a clear idea of what will happen.

Senator Dale Bumpers is against the new immigration program.

Newman said many of the inquiries he has received have been from European entrepreneurs. He said many have gone to Canada, but "the country of choice is the United States."

"The Canadian program has been a spectacular success," he said, "and we just sat here and let them do it."

Canada and Australia were also a lot easier to enter, said attorney Austin Fragomen of New York City. Even if all the uncertainties are cleared up, Fragomen said, "our program is not a competitive product" with the Canadian program.

Australia's program requires an investment of only $120,000. Canada's program requires $220,000 and no "hands-on" directorship of 10 employees. The U.S. requirements are obstacles, Fragomen said. "It's very difficult."

The other advantage with the Canadian program, said San Francisco investment banker Tony Angotti, is that Canada has a list of hundreds of pre-approved investments, so that when people invest, they know that after three, not five years, they can be citizens.

"People tend to forget that for the wealthy the United States is not the only game in town," he said. "If you are Hong Kong Chinese, wealthy and worried about 1997, you've been concerned for some time and you already have a New Zealand (or other) passport. These guys are the first ones out when trouble is coming," he said.

Irvin Philpot, office manager of a Palm Beach, Florida, holding company, said ads he has placed worldwide have generated substantial responses, "but we're not getting as many people that have a million in liquid cash that we had hoped for." To have that much, he said, "you've got to be worth between $10–$12 million."

Philpot, who has submitted four applications for the visas, said he was looking for investors in any of 27 companies, including fast-food franchises, light bulb manufacturing or taxi and limousine companies. The response has been cautious, he said.

7 GETTING INVOLVED

Critical Thinking

Pros and Cons Working with a group of students, consider the advantages and disadvantages of the "dollars-for-visas" program. List as many positive and negative aspects as you can. Then develop a plan of action that your group thinks should be taken on this issue. Present your findings to the class and be prepared to defend your group's position.

Decision Making

What Would They Say? What kind of person might be interested in coming to the United States under this program? Suppose you are such a person. Decide what country you are coming from, what business you are in, and so on. Write a letter in which you apply for admission to the United States as an investor. Include as much information about "yourself" as you can.

Critical Thinking

Create a Brochure So far, the dollars-for-visas program has not attracted many applicants. You are in the Public Relations department of the Immigration and Naturalization Service and your job is to create a brochure that will encourage people to enter the United States under this program. The brochure should be attractive, informative, and appealing to the type of person the program is aimed at.

8 FOCUS ON...
The Aging of America

The Administration on Aging (AOA) is one of the groups under the U.S. Department of Health and Human Services. The AOA's objective is to administer programs for the elderly. The AOA predicts that by the year 2030, 28% of U.S. citizens will be over the age of 60. How is the country going to deal with this dramatic change? Read the following article that describes a new campaign proposed by the Administration on Aging.

BEFORE YOU READ

Use the terms or phrases listed below to predict sentences that might possibly appear in the following article titled "Administration on Aging Announces The National Eldercare Campaign." Be sure you can defend the reasons why you feel that your sentences might possibly appear in this story. Share your sentences with a partner and see where you have similarities and differences.

Terms

older persons	organizations
independent	low-income
women	youth-oriented nation

WHILE YOU READ

Copy the graphic organizer below onto a sheet of paper. As you read the article, keep track of the goals and needs of the Administration on Aging's National Eldercare Campaign. After you have finished reading the article, decide for yourself if you think this campaign will work.

Administration on Aging's National Eldercare Campaign	
GOALS	NEEDS

123

Administration on Aging Announces The National Eldercare Campaign

By Joyce T. Berry
FROM: *AGING*
WINTER 1991

Within the next 40 years, the number of older persons in the United States will more than double. By the year 2030, the population of the entire country will look like Florida's, with 28 percent of Americans over the age of 60. One out of every four of us will be a senior citizen, and the number of people 85 and over will almost triple to 8 million.

The nation needs to start preparing for this dramatic change right now—with the same high energy and attention to detail that our military exhibited in planning and carrying out the rescue of Kuwait.

As an agency on "the front line" of the mission to reach elders at risk of losing their independence, the Administration on Aging [AOA] recently announced a new campaign. . . .

We are calling this nationwide multiyear effort the National Eldercare Campaign and its theme is "Eldercare: The Challenge of the 90's." The campaign has three major goals. The first is to use the media to raise public awareness of the implications for the individual and the society of the graying of America. We need to get citizens, corporations, and organizations to think about ways they can help today and in the future to meet the needs of the vulnerable elderly, who are the focus of the National Eldercare Campaign. This group includes elderly people who are ill, disabled, mentally impaired, abused or neglected, without a care giver to assist them when they need help, socially isolated, or living on a very low income. At special risk in the low-income group are women, minority individuals, and those living in rural areas.

The second goal of the campaign is to go beyond the traditional network of aging, social and religious organizations to broaden the base of support—in terms of both money and manpower—for development of in-home and community services for the elderly. We are asking businesses, educational organizations, foundations, and civic, neighborhood and youth groups to draw up their own Eldercare agendas—to consider what they can do for the elderly that isn't being done in their community or how they can improve services.

And third, the National Eldercare Campaign will spur the building of coalitions at the state and local levels to bring diverse organizations together to figure out solutions to the service needs of an older America.

Why the Need for This Campaign?

. . . Friendly visiting, store-to-door grocery services, adapting homes for the handicapped, home care help, chore work, emergency response systems for those who are ill and live alone—these are elements of the "supply line" that will protect our elders and keep them in the home they love.

Hundreds of community agencies now support or provide these services with federal and state funding. But there are gaps and weak spots in the "supply line," and towns, homes and apartments that it doesn't reach. In the city of Philadelphia alone, a recent survey found that 8,300 elderly residents said they needed home health services but weren't receiving them. If this is the situation in 1991, what will it be in 2010 or 2030?

With national and state budgets strained to the limit, there's a need to broaden the base of organizations and agencies involved in helping elders at risk. The aging network certainly can provide examples of various cities and towns that have already used this approach to address specific problems. . . .

8 GETTING INVOLVED

Critical Thinking

What Will It Be Like? Imagine that you are an elderly person in the year 2050. How old are you? What is your life like? Write a paragraph in which you describe your life–style. Tell about the health problems you have. Have any of the health problems from the 1990s been solved? What new problems are there? Try to be creative but reasonable as you imagine what life would be like.

Gathering Information

Compare Information The first paragraph of the article lists a number of statistics. For example: by the year 2030, 28% of Americans will be over the age of 60, and the number of people aged 85 and over will be 8 million. Go to your school or local library and find out how many people are currently over the age of 60. In addition, find out how many people are 85 and over. Then, using the figures given for the year 2030, calculate the percentage of people in each of these age categories. Make two graphs—one representing the current year and one representing the year 2030. Write a brief summary to go with your graphs.

Gathering Information

Interview an Expert Visit a residence for the aged in your town or city and spend at least two hours talking to the residents. Ask about what their lives are like, both good and bad. Share your experiences with the rest of the class in an oral or written report.

Human Populations

THINKING CRITICALLY ABOUT LOCAL ISSUES

You may want to work with a group of students for this activity. If so, make sure that each member of the group agrees about the issue that is chosen.

Prioritize the Issues Make a list of population-growth issues that are important to you and your community. Start by reviewing the brainstorming list that you made at the beginning of this module. If you make a scrapbook, or wrote in a journal, look over those as well. Also, review the magazine and newspaper articles in this module. Evaluate all the issues on your list. Then put them in order so the issue that has the highest priority for you or your group is at the top of the list.

Make the Issue Your Own Become an expert on your issue. Then decide what you as a citizen can do about the issue. Apply the process that you learned in the *S-T-S Problem Solving* module in the front of this book. Remember, the skills in the process are:

- Analyzing the Issue
- Gathering Information
- Making a Decision
- Planning Action

If you are working in a group, each group member has to contribute to all four stages. After completing the process, share your experience with your class.

CREATIVE THINKING

Suppose you are the leader of a country. Within the next ten years, the country's human population is expected to grow beyond what the country will be able to support. To keep the population from reaching that point, what would you recommend for education and ways to encourage couples to have fewer children?

Wrap-Up

HELP WANTED

Creating a Job Data Sheet Join with your classmates to make a list of all the careers mentioned in the articles of this module. Then form a group with two or three of your classmates. Each group in the class should choose a different career. Prepare a data sheet for your group's career. At the top of the sheet, write the name of the career. Then write the following headings. Leave space after each heading to record information.

- Typical Duties and Responsibilities
- Type of Work Environment
- Salary Range
- Educational Requirements
- Other Information
- Sources for More Information

You can refer to the Community Resource Directory for more information about careers.

HUMAN HEALTH and Disease

> "All interest in disease and death is only another expression of interest in life."
> —Thomas Mann, 1924

HUMAN HEALTH and Disease

IN THIS MODULE

Overview	131
Discovery	138
1 "U.S. Reorganizes Nutrition Advice" from *The New York Times*	140
2 "Schools Relax TB Deadline, Let Kids In" from *Newsday*	145
3 "Doctors' Group Urges Tough Laws on Smoking" from *The Los Angeles Times*	148
4 "Senate Panel Hears of Pesticide Harm Abroad" from *The Los Angeles Times*	152
5 "Better Safe Than Sorry?" from *Time*	156
6 "HIV Tests in the Health Profession" from *The Washington Post*	160
7 "Behind Animal Testing Is Profit Motive" from *Roanoke Times & World News*	164
"Think Animal Testing Is Inhumane? Explain It to the Sick, Dying" from *Charlotte Observer*	166
8 "Laying Siege to a Deadly Gene" from *Time*	169
9 "Group, Legislations Targets Women's Health Research" from *The Detroit News*	173
Wrap-Up	176

HEALTHY LIVING

In the United States over the next 12 months, 3 out of every 4 members of your class will get a cold. In the same year, nearly 20 million people will undergo some type of surgery. Roughly the same number will be treated for back problems. One in ten people will get an ulcer. One-half million people will suffer a stroke and nearly a million will be diagnosed as having cancer.

It certainly seems from these statistics that good health is hard to maintain. During our lifetimes, we are all stricken with illnesses. In many instances, we treat the illness and get better. Other times, the illness cannot be treated and our bodies are permanently damaged. Unfortunately, many of us take our good health for granted. We don't appreciate our physical, mental, and emotional well-being until we're not well.

Good health means more than just not being sick. Good health includes **physical health, mental health,** and **social health.** Physical health is the health of your body and its functions. Mental health is the health of your mind, including a positive self-image. Social health is the way you deal with other people and your relationships with them. Another word for good physical, mental, and social health or well-being is **wellness.**

Some activities, such as walking on thin ice, are always dangerous. Other activities can be made safe by wearing protective gear.

Becoming Unhealthy

Nobody wants to become unhealthy, but unfortunately we all become ill at one time or another. There are many different reasons why people become unhealthy. About nine million people in the United States are seriously injured due to accidents each year. In the student population, accidents are the number one cause of death. About one-third of these accidents occur in the home.

Infectious Diseases

Diseases are generally classified into two groups, infectious diseases and noninfectious diseases. An **infectious disease** is caused by a **pathogen,** or microscopic organism, that enters the body. Bacteria and viruses are the most common pathogens.

How do pathogens actually cause disease? Usually by taking over healthy cells and interfering with their life processes. Bacteria accomplish this by releasing **toxins,** or poisons, into the cells. Viruses directly invade healthy cells and multiply within them until the cells are destroyed.

Certain infectious diseases are transmitted or spread from person to person. Such diseases are called **contagious** or communicable diseases. Some examples of infectious diseases are chicken pox, tuberculosis, measles, hepatitis, AIDS, and influenza (the flu).

One infectious disease that you most likely have experienced is the common cold. More than 100 kinds of viruses can cause this contagious disease. The virus is spread through contact with an infected person. When an infected person sneezes or coughs, droplets of mucus containing the virus are released into the air. When you touch something

COMMON INFECTIOUS DISEASES AND HOW THEY ARE SPREAD

Disease	How Disease Is Spread
Chicken pox	contact with infected person contaminated droplets in the air
Common cold	contact with infected person contaminated droplets in the air
German measles	contact with infected person contaminated droplets in the air
Influenza	contact with infected person contaminated droplets in the air
Syphilis	sexual contact
Tuberculosis	droplets in the air
Hepatitis	contact with infected person transfusion of contaminated blood
AIDS	direct contact with infected blood or body fluids use of contaminated needle sexual contact

that has the virus on it, or inhale one of those droplets, the virus enters your body, and you may develop a cold.

In the United States, a cure for many infectious diseases is available. As an infant or a small child, you were given shots or immunizations against many infectious diseases, such as German measles and polio. In many other countries, especially developing countries, the immunizations for many infectious diseases are often not available at all or are available only in limited quantities because of widespread poverty. As a result, many more people die from infectious diseases.

AIDS

One of the most deadly infectious diseases is **AIDS**. AIDS stands for **A**cquired **I**mmune **D**eficiency **S**yndrome. AIDS is a disease of the immune system. It is caused by the **h**uman **i**mmuno**d**eficiency **v**irus (HIV). HIV cells destroy white blood cells that are normally used to fight off disease. As a result, a person with AIDS cannot destroy pathogens that enter his or her body. A person who has AIDS dies not from AIDS—which is a syndrome (a set of characteristics)—but from complications or diseases that his or her body could not fight off.

AIDS is generally spread in four ways. One way is by having sexual relations with an infected person. A second way is by sharing hypodermic needles, as some drug users do. A third way is by coming in contact with the blood or bodily fluid of an infected person. Lastly, an unborn fetus can develop AIDS if the HIV virus is passed on to the fetus.

Although there is no known cure for AIDS, it can be prevented. By not engaging in any of the "high risk" behaviors listed above, you can prevent yourself from getting AIDS. Abstinence is the only "no risk" sexual behavior. In addition, learning as much as you can about the disease and how it is spread will help.

Noninfectious Diseases

Any disease that is not caused by pathogens is a **noninfectious disease.** Cancer, cardiovascular (heart) diseases, arthritis, and diabetes are all noninfectious diseases. Because these diseases are not caused by pathogens, they are not contagious.

Noninfectious diseases are caused by a variety of factors. Many cardiovascular diseases are the result of the breakdown of tissues and/or organs of the circulatory system. Cardiovascular diseases, which include arteriosclerosis (ahr-tihr-ee-oh-sklu-ROH-sis), or hardening of the arteries, high blood pressure, heart attacks, and strokes, are the leading cause of death in the United States among both men and women.

Cancer is a noninfectious disease that is caused by the uncontrolled growth of abnormal cells. Cancer cells reproduce rapidly in a disorganized way. As these cells grow, they compete with healthy cells for food and oxygen. In the process, the cancer cells destroy normal body cells, tissues, and organs.

Cancer can occur almost anywhere in the body. Many types of cancer are fatal, but some types can be cured if they are caught early enough. Among the **carcinogens,** things that can cause cancer, are pesticides, cigarette smoke, ultraviolet light, radon, some food preservatives, and wastes from coal- and oil-burning industries.

Maintaining Wellness

With the thousands of ways a person can be injured and thousands of differ-

Exercise is important in maintaining good health. It makes your body stronger, promotes better posture, and improves endurance.

ent kinds of diseases waiting to attack the body, how can one possibly stay healthy? As already mentioned, "unhealthiness" is a fact of life. However, there ARE some things you can do to stay healthy.

Good Nutrition—Have you ever heard the expression "you are what you eat"? There is a lot of truth in that statement. The foods you eat help determine how you look and feel. A **balanced diet,** one that includes foods from each of the different food groups, supplies the body with the **nutrients** it needs to run properly. Nutrients are substances found in foods that the body needs for growth, energy, and the life processes, such as respiration and circulation. Vitamins, minerals, carbohydrates, proteins, and fat are examples of nutrients.

Not eating properly can lead to disease. Your body needs the nutrients that each type of food provides. If you do not eat a balanced diet, you may become the victim of a **deficiency disease.**

In developing countries, deficiency diseases are common. You may have seen photos of small children with bloated stomachs. These children are suffering from poor nutrition and deficiency diseases.

In addition to keeping your body fit today, a balanced diet may reduce the chance that you will be stricken with certain diseases in the future. Both the American Cancer Association and the American Heart Association suggest that a diet low in fat and high in fiber helps promote good health. Their recommendations are based on studies that show that diets high in sodium, saturated fat, and cholesterol increase the risk of heart disease and certain types of cancer.

Unfortunately, the typical diet in the United States has a higher fat content than the typical diet in any other country. Many scientists link this high-fat diet to the large number of Americans that suffer from heart disease, obesity, and cancer of the colon, breast, and uterus. Studies on the diets of other countries bear witness to this link. For example, the traditional diet in Japan is low in fat and cholesterol, and the death rate from heart disease is low there. However, studies show that when Japanese migrate to the United States and adopt an "American" diet, their death rate from heart disease rises.

Exercise—Exercise is also important in maintaining good health. Studies have shown that routine exercise lowers the risk of heart attack. This is because physical conditioning increases the ability of the heart to pump blood throughout the body. Since the heart can pump more blood with each beat, it doesn't have to beat as often, which results in a lower heart rate. People with lower heart rates have fewer heart attacks than other people. In addition, physically active people have fewer irregularities in heart rhythm, lower blood pressure, and reduced levels of blood fats.

Mental and Social Health—The amount of stress in a person's life also affects a person's health. Any demand you put on your body is stress. There is both good stress and bad stress. Getting an "A" on a test is an example of good stress. Trying to finish all of your homework in homeroom is an example of bad stress.

While stress is a part of everyone's life, certain lifestyles increase its presence. Constantly "spreading yourself too thin" by attempting to do too many things increases stress. Studies have shown that high-stress lifestyles can contribute to such health problems as

Talking through your problems with friends, as well as adults, can help reduce stress.

headaches, rashes, high blood pressure, and ulcers, and can even increase the likelihood of a heart attack. How do you think you can reduce the amount of bad stress in your life?

There are many ways to deal with stress. Some of the most effective ways include exercise, yoga, talking with others, and relaxation techniques. Everyone deals with stress differently. What works for you may not work for a friend. The important thing to remember is to deal with your stress. If you cannot deal with your stress, you should talk to a parent, teacher, counselor, or someone who may be able to help you.

Lifestyle Choices

Choices you make about what you put into your body affect your ability to stay healthy. For example, by eating balanced meals, exercising regularly, and maintaining positive self-esteem, you will stand a better chance of living a longer and healthier life. Other choices you make, such as smoking cigarettes, may shorten your life span.

Tobacco—One of the main things you can do to stay healthy is NOT smoke cigarettes! About one-fourth of heart attack deaths result from smoking. One-fifth of all cancer cases are linked to cigarette smoking. Smoking cigars and pipes and chewing tobacco are just as harmful as smoking cigarettes. The use of tobacco increases the likelihood of high blood pressure, high cholesterol, coronary heart disease, respiratory ailments such as bronchitis and emphysema (em-fih-SEE-muh), peptic ulcers, hypertension, and even sleep problems!

Drugs—Any chemical substance you put into your body is a **drug.** Drugs can be helpful in the treatment of disease. However, many people misuse and abuse drugs.

Drug abusers become physically and emotionally dependent on a drug. This means the body gets used to the drug. The drug user needs stronger and stronger dosages to get the same effect. Drug abuse often leads to overdose and even death.

Drug abuse is a problem in the suburbs as well as in big cities. It is probably one of the biggest problems society faces today. In order to get money to pay for drugs, some drug users resort to stealing and violence. A large number of violent crimes today are drug related.

Drugs affect both the people who take them and society in general.

How do you think society should deal with the problems related to drugs?

Alcohol—The most widely abused drug is alcohol. Alcohol is responsible for thousands of traffic accidents and deaths in this country. Alcohol slows down your reaction time and impairs your judgment. Many states have strict drunk-driving laws. Groups such as MADD (Mothers Against Drunk Driving) and SADD (Students Against Drunk Driving) are pressuring politicians to enact and enforce even stricter laws.

Making choices about drugs, alcohol, and tobacco can affect your health in the future. Your body takes a long time to heal once it has been damaged by drugs, alcohol, or tobacco.

Environmental Problems

There are some things, such as environmental problems, that affect your health that you have less control over than your diet, alcohol, and drugs. A side effect of the modern technology that has made our lives easier is a string of environmental health hazards. You probably take for granted that the tap water in your home is pure. Yet, nearly 10,000 people in the United States are stricken with infectious diseases each year by drinking tap water. Studies have shown that the water supply in thousands of places in the nation contains potentially harmful substances such as pesticides, lead, asbestos, viruses, and organic chemicals known to cause cancer.

The air you breathe can also be potentially harmful to your health. Industrial waste products add pollutants to the air that can worsen such respiratory diseases as bronchitis and asthma. Breathing polluted air may also increase the risk of being stricken with emphysema and cancer.

Even the noise that surrounds you each day can negatively affect your health! Millions of people in the United States have suffered hearing loss because of environmental noise. Most of these injuries are due to continual exposure to loud noise, which damages the inner ear. In addition to noise produced outside by such things as airplanes and heavy machinery, loud noise occurs inside as well. Playing your stereo or television too loud damages your ears also. Noise from stereo headphones is especially damaging to ears.

Excessive noise can also affect a person's mental well-being. Studies have shown that excessive noise makes people feel tired, nervous, and irritable. It can even affect the way a person interacts with other people.

Health and Society

You have probably concluded by now that you can increase your likelihood of staying healthy by making careful decisions about your lifestyle. Scientific study has provided you with a wealth of information about the causes of disease. You can use this information to make careful, intelligent decisions about the way you live.

Yet, in spite of the facts at hand, some members of society choose to pursue an unhealthy lifestyle that increases their risk of disease. At the present time, nearly 55 million people in the United States continue to smoke cigarettes knowing very well that they are shortening their life expectancy. Some people argue that government has a responsibility to force such individuals into kicking their habit and leading a more healthy life. Others argue that each individual has the right to lead whatever type of lifestyle he or she chooses. What do you think?

DISCOVERY

HOW CAN YOU MODEL THE SPREAD OF DISEASE?

Many human diseases, such as the common cold and other contagious diseases, spread through direct contact between people. To gain an understanding of how quickly such a disease can spread, try the following activity.

Materials (per class)

numbered adhesive labels
paper and pencil

Procedure

1. Your teacher will give you an adhesive label with a number on it. Attach the label to the front of your body where it is clearly visible.
2. On a sheet of paper, make a table like the one shown.
3. Shake hands with a classmate. Record his or her number in row 1 of your data table.
4. Shake hands with another classmate. Record his or her number in row 2 of the table.
5. Shake hands with a third classmate and record his or her number in row 3 of the table.
6. Your teacher will randomly select a number and announce it to the class. Imagine that the person wearing that number has a disease that is transmitted through physical contact.
7. If you are the person with the disease, check your data table to identify the three people with whom you shook hands. Notify those people that they are now also infected with the disease.
8. If you are one of the three people who just learned that you were infected, check your data table. Identify all the people whose hands you shook after you became infected. Notify these people that they are now also infected.
9. If someone in Step 8 has just informed you that you are infected, check your data table to see whose hand you shook after you became infected. Notify these people that they, too, are infected.
10. If you became infected at any point during this activity, write your number on the chalkboard.

DATA TABLE	
Handshake Round	Identification number of person
1	
2	
3	

Observations

1. How many people in the class were infected? What percent of the class does this represent?
2. To get a disease, was it necessary to have had direct contact with the first person who had the disease? Explain your answer.
3. Do your chances of becoming infected with the disease increase or decrease with each round of handshaking? Explain your answer.

Conclusions

1. What would happen with the spread of the disease if the rounds of handshaking continued on for a few more rounds?
2. Each time you shook hands, you might have been shaking hands with a person who had a disease. How is this similar to the spreading of diseases in real life?
3. Make a list of all the diseases that you know that are transmitted by physical contact.
4. Suppose the activity you just completed had been conducted in the following way: An infected person shook hands with two people. Then, those two people each shook hands with two new people. Then, each of those people shook hands with two new individuals. How many people would be infected at that point? How many people would be infected after one more round of handshaking? Make a diagram to illustrate this process of infection.

BRAINSTORMING

Collaborative Learning What are some problems or issues that relate to staying healthy and preventing disease? What are some issues concerning the spread of infectious diseases? Spend at least five minutes with a group of your classmates making a list of health and health-related issues. Express each issue as a question. For example, should smoking be prohibited in all public areas? Should AIDS testing be required for all health-care workers? Remember, when you brainstorm, do not criticize each other's ideas. After your group has completed the list, have one group member report to the class. How many issues did the class list? Keep a copy of your brainstorming list in your notebook. You will need it for some of the activities later in the module. Guidelines for brainstorming can be found in the skill lesson on page 231.

SCRAPBOOK

Gathering Information Clip out advertisements for health and health-related products and services from magazines and newspapers. Classify the advertisements into those for products, such as vitamin supplements or cold remedies, and those for services, such as health clubs or rehabilitation centers. Indicate the age group(s)—young child, teenager, middle-age adult, the aged—for which you think each advertisement is intended. Which of the products or services appeal to you? Continue adding to your scrapbook as long as your class works on this module. See the skill lesson on page 232 if you need help getting started.

JOURNAL WRITING

Critical Thinking In your journal, identify the diseases or other health related concerns that you think would most threaten the health of someone in a developing country. Indicate how these factors compare to the factors that are likely to affect your own health. Express your thoughts about the health needs of people in different parts of the world. What would you do to improve people's health in your own country and in developing countries? During the time you work on this module, continue to add to your journal. If you need some ideas on writing in a journal, see the skill lesson on page 233.

1 FOCUS ON...
Industry Pressure vs. Public Health

The United States Department of Agriculture (USDA) is a government agency that has two major objectives—to promote agricultural products, such as meat and dairy products, and public health education. These two objectives are often in conflict. In April 1991, the USDA introduced an "eating right pyramid" that suggested that people eat more bread and grain products and less meat and dairy products. Pressure from the meat and dairy industries made the USDA withdraw its pyramid. One year later, the pyramid was adopted. Was it the right decision to delay the adoption of the "eating right" pyramid?

BEFORE YOU READ

Work with three other students to develop a graphic organizer that illustrates what you already know about good nutrition. Discuss within your group what kinds of information you want to include and how you would arrange the information. Add as much information as you can. After you have read the article, you may want to go back and add additional information to your graphic.

WHILE YOU READ

As you read this article, you will find a number of different viewpoints or positions, regarding the adoption of the food pyramid. Find all the players that hold these different views. Complete a chart like the one shown listing each player's name, position, and role regarding the food pyramid.

FOOD PYRAMID		
Player	Role	Position

U.S. Reorganizes Nutrition Advice

Food educators win battle to depict 5 basic groups in a pyramid design

By Marian Burros
FROM: THE NEW YORK TIMES
APRIL 28, 1992

Washington, April 27—After a year and almost $1 million spent on debating how best to teach the public about nutrition, the Agriculture Department on Tuesday will adopt as its primary educational device a pyramid that divides foods into five groups.

The Food Guide Pyramid will replace the venerable [impressive on account of age] four food groups, which have been used to teach nutrition in schools since the 1950's. Its publication will affect the way children learn about nutrition for years to come.

The pyramid was to have been adopted a year ago, but Edward Madigan, the Agriculture Secretary, delayed its publication because of questions about its effectiveness as a teaching aid and opposition from the meat and dairy industries. It advises the public to use fats and oils "sparingly" and says that meats and dairy products should be eaten less frequently than fruits, vegetables and grains.

Problems With Pyramid

Agriculture Department staff members say Mr. Madigan was not pleased with the choice of the pyramid over a bowl-type illustration, which the meat and dairy industries felt did not appear to rank the food groups. "The political people were forced into this decision by the internal staffers, the Department of Health and Human Services and the professional community," said one department staffer who said she would speak only on condition of anonymity because she feared she might otherwise lose her job.

Mr. Madigan denied that he had been pressured into his decision. Speaking of two assistant secretaries in his office, Mr. Madigan said, "They came to me with their conclusions and the reason why the pyramid was superior and I accepted that."

The original pyramid, with minor modifications, was chosen because it was considered the best way to convey [communicate] the Government's most important nutritional messages: that Americans need to eat less fat and eat a certain amount of food each day from each of the five food groups.

The pyramid places the grains group at its base, vegetables and fruits, now two groups instead of one, just above it

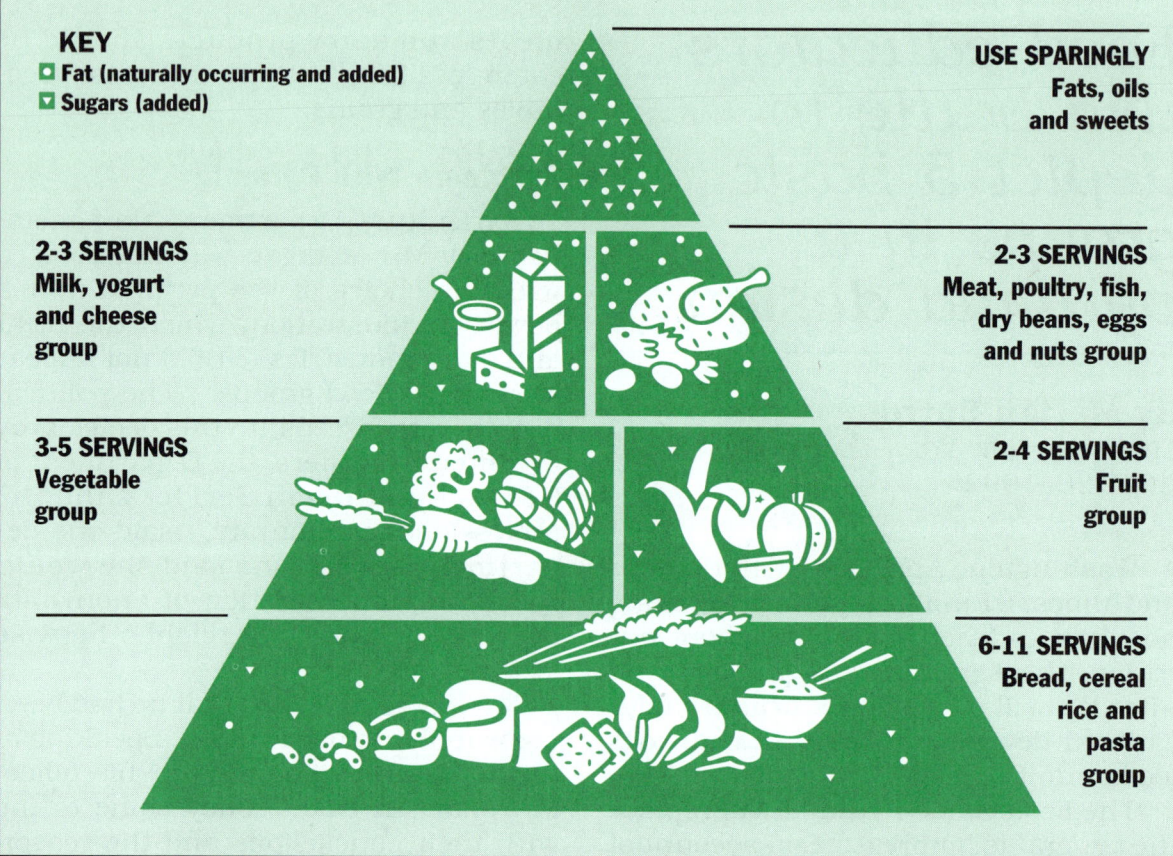

NOW, 5 FOOD GROUPS
The Federal Government has adapted the Food Guide Pyramid as its primary devise for educating the public about nutrition.

KEY
☐ Fat (naturally occurring and added)
▽ Sugars (added)

USE SPARINGLY
Fats, oils and sweets

2-3 SERVINGS
Milk, yogurt and cheese group

2-3 SERVINGS
Meat, poultry, fish, dry beans, eggs and nuts group

3-5 SERVINGS
Vegetable group

2-4 SERVINGS
Fruit group

6-11 SERVINGS
Bread, cereal rice and pasta group

Adapted from: The New York Times

Source: Department of Agriculture

and a narrow band of meat and dairy products near the top. At the apex [point] are fats, oils and sweets, which are not considered a food group, with the note to "use sparingly."

The placement of the meat group right under the fats and oils angered the National Cattlemen's Association, which asked Mr. Madigan to withhold publication of the pyramid in April 1991, just days before it was to be released.

The dairy industry joined in the protest. Both groups felt that the placement of their products in the pyramid suggested an undesirable ranking. Now they seem resigned to the choice of the pyramid, which was shown last year to more than 3,000 children and adults and their reactions noted.

When the National Cattlemen's Association was in Washington for its annual meeting in March its board met at the Agriculture Department with Steve Abrams, the deputy assistant secretary for food and consumer services. He assured them that ordinary consumers do not see the food groups in the pyramid as good or bad.

"They did not see any ranking by consumers," said Mark Armentrout, chairman of the food policy committee of the National Cattlemen's Association.

'Good Food and Bad Food'

Karl Hoyle, a spokesman for the National Milk Producers' Federation, said: "We would have preferred the bowl. We can certainly live with [the] pyramid, but it still gives the image of good food and bad food."

The nutrition professionals inside and outside the Agriculture Department feel vindicated, too. "The people in the field should feel proud they have some muscle," the department staff member said. "The nutrition community stood up and said enough is enough. The political people wanted to drop it and said it would be a one-day story, but it just didn't die. The research would never have been done if it hadn't been for the pressure. The research was based on sound scientific rationale instead of political gains. When the results came out it was so clear cut that they could not manipulate it."

The staff member [said] that Mr. Madigan's office still concluded that the bowl was the best choice.

The professional staff and the outside consultants could not agree with the conclusions of the political staff. "The internal people and the outside consultants were extremely consistent in their conclusions," the staff member said. Conversations with members of the internal advisory panel and outside advisors confirm this view.

Aid in Teaching

"Looking at the question as an educator, there is no question which one I would pick," said Cheryl Achterberg, an associate professor of nutrition at Penn State and one of the outside consultants. "The pyramid depicts the concepts more clearly and is more teachable."

Mr. Madigan denied that he had tried to disregard the data. "I don't know where you're getting this stuff, but it goes beyond anything I have been involved in," he said.

> **"** *The research was based on sound scientific rationale.* **"**

Some nutritionists are outraged at the amount of money spent to confirm that the pyramid was the proper choice.

"I'm appalled they wasted almost $1 million of taxpayer money to prove what earlier research had already shown," said Bonnie Liebman, director of nutrition at the Center for Science in the Public Interest, a consumer advocacy group.

A copy of the Food Guide Pyramid is available at no charge from the United States Department of Agriculture, Washington, D.C. 20250.

1 GETTING INVOLVED

Decision Making

Plan a Balanced Diet For one day, make a list of all the food and beverages you consume. Include approximate amounts, such as one apple, bowl of cereal with milk, and so on. Using the "Food Guide Pyramid" pictured in the article, compare what you ate to the guidelines presented in the pyramid. Were there any types of food of which you ate too much? Too little? Plan a diet for one day based on the pyramid. Try to follow that diet. Was it easy or hard to follow the eating right guidelines? Write a brief summary of your diet and how easy or difficult it was to follow and why.

Cooperative Learning

Set up a Nutrition Awareness Fair As a class, set up a nutrition awareness fair. Divide the class into groups. Each group should have a different responsibility. One group may wish to write to the United States Department of Agriculture to obtain copies of the Food Guide Pyramid to distribute throughout the fair. Other groups may want to gather additional information by writing or calling a few of the organizations found in the General Organizations Section of the Community Resource Directory starting on page 244.) Tell the various organizations what you are doing and request that they send you brochures, posters, and any other information. Another group of students may want to arrange for a speaker from the local chapter of the American Cancer Society or National Dairy Council. Another group of students may want to hang posters throughout the school advertising the fair. For some hints on getting started see page 243.

Critical Thinking

What Is Their Position? For each of the players listed below, describe what you think their position is. You may wish to organize your information in a chart similar to the one shown. First, state the issue in the form of a question. Then, write the names of the player. Next, describe the position of each player. Some of the players listed are not mentioned in the article, so you will have to infer, or interpret, their position from their jobs.

Players
1. milk producer
2. cattle rancher
3. American Heart Association member
4. nutritionist
5. Secretary of Agriculture
6. fruit/vegetable stand owner

ISSUE:	
Player	Position

2 FOCUS ON...
The Possible Spread of Tuberculosis

Tuberculosis, or TB, is a contagious, infectious disease caused by bacteria. It passes from one person to another very easily. Tuberculosis is common in poor areas of cities. In order to reduce the spread of TB, New York City requires all new students to be tested for TB. On the first day of school in 1991, school administrators discovered that hundreds of new students had not been tested. Rather than keep those students from entering school, the school board extended the deadline for testing and allowed the students to enter the school system. Was this an appropriate action? Why or why not? Read the article and see what you think.

BEFORE YOU READ

Below is a list of health situations. Copy the list onto a separate sheet of paper. Next to each statement, write either "go to school" or "stay at home." Some situations can be placed in either category so write a short explanation about the reasons for your decisions.

Situations

1. You have a sore throat.
2. You just got the measles.
3. You didn't have a flu shot.
4. You have a cold and a fever.
5. You have not yet been tested for tuberculosis (TB).
6. You have a bad cough.
7. You have an upset stomach.

WHILE YOU READ

Copy the chart shown onto a separate sheet of paper. As you read this article, fill in the chart with the different players and their beliefs. When you have finished reading the article, complete the following belief statement:

The belief that is closest to my own is _____.

PLAYER	BELIEF

Schools Relax TB Deadline, Let Kids In

By Kevin McCoy
FROM: *NEWSDAY*
OCTOBER 3, 1991

Struggling to cope with hundreds of school children who failed to get mandated tuberculosis tests, city health and education officials cancelled an October 1 [1991] screening deadline, officials acknowledged yesterday.

The move averted the threat that hundreds of students around the city would be barred, at least temporarily, from attending classes. But it also ignited a debate over whether the testing delay would increase the growing spread of the dangerous disease, which is contracted by breathing air exhaled by an infected person.

City health officials yesterday insisted that the decision to give parents and guardians an indefinite extension of the testing deadline would not pose a serious health threat.

"It does not pose an inordinate risk to delay the testing deadline," said Margaret Karanjai, a spokeswoman for the city Department of Health. "The greater danger would be having these children out of school."

But a spokesman for the union representing most city-employed physicians charged that allowing the children who have not been tested for tuberculosis to remain in classes "puts students at risk."

"Thousands of school children who may have the disease may go undiagnosed or untreated for weeks or months. That's the very real danger here," said the union official, Dr. Donald Meyer, executive director of the Doctors Council.

The debate revolves around a new city requirement that all new public, private and parochial school students be tested for tuberculosis within 14 days of the start of school. Health officials imposed the requirement after 1989 tests among nearly 3,000 needy school children showed that 6.8 percent had been exposed to tuberculosis.

Citywide, 146 cases of tuberculosis were diagnosed in 1990 among children under age 14, nearly double the 74 cases in 1989. New York City has reported 2,221 cases of the disease among children and adults so far in 1991, compared with 2,492 cases at this time last year.

Tuberculosis, a disease generally transmitted through the coughs of an infected carrier, requires months of drug therapy and can be fatal if not treated.

Despite the city's testing mandate, hundreds of the estimated 25,000 native born and immigrant students entering New York schools for the first time this year were not tested by the city's October 1 screening deadline.

"We couldn't have these children not being in school. That would be a major problem," said James Vlasto, a spokesman for city schools Chancellor Joseph Fernandez.

Instead, city education and health officials this week decided to require parents and guardians to produce written proof that their children have appointments to be tested by a private doctor or at a city clinic.

Karanjai yesterday said the city has identified 130 public and private testing sites. The department is also arranging tuberculosis screening sites and dates in each city school district, Karanjai said.

Meyer said private hospitals and city clinics are already overwhelmed with students needing tests.

2 GETTING INVOLVED

Gathering Information

Interview an Expert Work with three other students to prepare 5 to 10 questions concerning tuberculosis. You may wish to include questions that focus on the history of TB, the inoculation process, and the present danger of school children's contracting tuberculosis. In addition, find out what your school's policy is regarding TB. Interview either the school nurse or someone in your city or county health department. (You can find the addresses and phone numbers in the blue pages of your telephone directory.) Write a report on your findings.

Critical Thinking

Find Out More Go to your school or local library and research the number of cases of tuberculosis in different countries. The *World Almanac* would be a good source to use. Make a database including such information as: continent, country name, number of cases in each of the last five years, and any other relevant information. Review your database and consider why some countries have a higher incidence of TB than other countries. How does the number of TB cases change from continent to continent? Present your database to the rest of the class.

Decision Making

What Do You Think? Answer the following questions in complete sentences. Remember, there are no right or wrong answers—these are your belief statements.

1. Should students who haven't been tested for TB be allowed to enter school?
2. Who do you think should cover the cost of testing?
3. Who should pay for treatment if a TB test comes back positive?
4. If a student testing negative for TB contracted TB from an untested student, who should be financially responsible for the treatment: the school board? the parents of the infected student? Why?
5. Who should pay for the testing of poor children?

3 FOCUS ON...
Cigarette Vending Machine Ban

It is currently against the law for a store owner to sell cigarettes to youths under the age of 18. However, public vending machines cannot be regulated in the same way. For this reason, the Los Angeles County Medical Association has proposed that cigarette vending machines be banned. In addition, the proposal includes prohibiting smoking in restaurants and bars. The tobacco industry, as well as the California Restaurant Association, are against this ban. Is banning cigarette vending machines a good way to deal with the problem of underage smoking? Read the article and decide for yourself.

BEFORE YOU READ

Below is a list of statements that deal with smoking issues. Which of the statements do you agree with? Write these statements on a sheet of paper. For those statements that you do not agree with, rewrite them so that you can agree with them.

Statements

1. No one should be allowed to smoke in a public place.
2. Nonsmokers should not have to sit next to smokers in any public place.
3. No one under the age of 18 should be permitted to buy cigarettes anywhere.
4. Tobacco companies should not be permitted to advertise in any magazine targeted for youths.
5. Cigarette vending machines should be banned.
6. People should be allowed to do whatever they wish in a free country as long as they do not endanger others.

WHILE YOU READ

Copy the chart shown below onto a separate sheet of paper. First, state the issue in the form of a question. As you read this article, fill in the chart with the different players and their beliefs on the issue. Include your own belief as well.

ISSUE:	
Player	Belief
Los Angeles County Medical Association	
Tobacco Institute	
California Restaurant Association	
You	

Doctors' Group Urges Tough Laws on Smoking

Statewide ban on cigarette vending machines . . . asked by county medical association

By Irene Wielawski
FROM: THE LOS ANGELES TIMES
MARCH 7, 1991

The Los Angeles County Medical Association Wednesday proposed tough anti-smoking legislation that would make California the country's most inhospitable [intolerant] state to smokers.

The medical association wants statewide bans on cigarette sales through vending machines, and on smoking in restaurants, bars and hospitals. The measures, according to association President David Chernoff, are aimed at making it harder for youngsters to take up smoking, and at reducing health hazards from so-called passive smoking—the inhalation of smoke from the cigarettes of others.

The group expects swift endorsement [approval] of the measures next week by the California Medical Association at its annual delegates meeting, which would bring the full muscle of organized medicine in California to make the proposals law.

But as news of the doctors' plan spread Wednesday, so did industry opposition.

"We, of course, will forcefully oppose it," said Walker Merryman, spokesman for the Tobacco Institute in Washington, the trade association [association that works for the promotion of certain goods] for cigarette manufacturers. Merryman said he wasn't surprised by such a move in California. . . .

Such rules have served primarily to hurt the restaurant and bar business, Merryman said, without evident improvement in patrons' health. Adoption

> ❝*The law is saying that second-hand smoke is terrible, but only in restaurants.*❞

of the medical association's proposal would make California the most restrictive state in the country, he added. . . .

The California Restaurant Association, meanwhile, said it will oppose the measure because the proposed smoking bans target restaurants and bars, rather than all public places. . . .

"It is sending a very mixed message to the public; it is saying second-hand smoke is terrible but only in restaurants," said Jo-Linda Thompson, lawyer for the association, adding that the medical association should have consulted with her group before going public with the proposal.

"I certainly have spent more time in airports with people smoking around me than I have in restaurants," Thompson said. "If it is a carcinogen [cancer-causing substance], then it should be prohibited in all public places: bowling alleys, shoe stores, video parlors, any public gathering place." . . .

The real reason dinosaurs became extinct

A ban on vending machine cigarette sales would cut off a major source of cigarettes to children, Chernoff said. Surveys indicate that 90% of adult smokers began their habit as children or adolescents. Although California and 36 other states prohibit sales of cigarettes to people younger than 18, vending machines in public places permit children to circumvent [get around] that restriction. . . .

Two California cities, Santa Monica and Rancho Mirage, recently banned vending machine cigarette sales. Los Angeles and West Hollywood are considering similar bans. The Los Angeles City Council last year defeated a ban on smoking in restaurants.

Assemblyman Bruce Bronzan (D-Fresno), who is chairman of the Assembly's Health Committee and an outspoken foe [opponent] of smoking, said he welcomed the medical association's efforts, but hoped the group would lend support to anti-smoking bills already before the Legislature.

Among those are measures to increase the tobacco tax by 10 cents a pack, permit cigarette vending machines only in bars or other places inaccessible to children, ban the free distribution of tobacco products and require licensing of tobacco sellers so that they could be penalized—like liquor sellers—for selling cigarettes or other tobacco products to minors.

3 GETTING INVOLVED

Critical Thinking

Make a Persuasive Speech A persuasive speech is a speech that tries to influence the attitudes, beliefs, or behavior of an audience. You make a persuasive speech to change people's minds or to move people to take some action. Make a 5-minute persuasive speech on your feelings about the issue of banning cigarette vending machines. For some helpful hints on making a persuasive speech, see page 240.

Cooperative Learning

Discuss the Issue Get together with five other classmates. Have each person take one of the following roles to play:
1. moderator
2. smoker
3. vending machine distributor
4. restaurant owner
5. nonsmoker
6. Tobacco Industry representative

Using the facts found in this article, as well as your own opinion, prepare a one-minute statement concerning the issue raised in the article. After each opening statement has been made, each player may ask a question of each of the other players. At the end of the question-and-answer session, the moderator should summarize what was said, and decide who had the most convincing argument.

Gathering Information

Take A Public Opinion Survey A public opinion survey is a tool used to gather information about people's attitudes. In a group of four students, develop a public opinion survey to find out what the public thinks about banning cigarette vending machines.

Develop at least five questions. After developing the questions, each member of the group should take a different age group to ask the questions of (this is your target audience). Compare and analyze the results from the surveys. Is there a difference in how certain age groups feel about this issue? Summarize the group's findings on a graph and in a written summary.

FOCUS ON...
4 Pesticide Use Abroad

Pesticide use has increased in many Central American countries in order to raise the amount of their production of fruits and vegetables for export. However, studies on certain pesticides have shown them to be harmful. The use of these harmful pesticides has been banned in the United States. However, chemical companies in the United States are exporting these pesticides to Central American countries where there are no restrictions. The farmers in these countries are now asking the U.S. Congress to ban the export of these potentially dangerous pesticides. What is more important, free enterprise (profit) or human health? Read this article and decide for yourself.

BEFORE YOU READ

Use the words listed below to predict sentences that might possibly appear in the article titled "Senate Panel Hears of Pesticide Harm Abroad." You may use more than one of the words in your sentence. Be sure you can defend the reasons why you feel that your sentence might possibly appear in this story. Share your sentences with those of a partner and see where you have similarities and differences.

Words
chemical
farmers
hazardous
pesticide
Senate Agricultural Committee

WHILE YOU READ

Copy the graphic organizer shown onto a separate sheet of paper. As you read the article, fill in the graphic organizer with the different players and each player's belief statement. After you have completed the article decide whether or not you support the Circle of Poison Prevention Act, and complete the sentence under the graphic organizer.

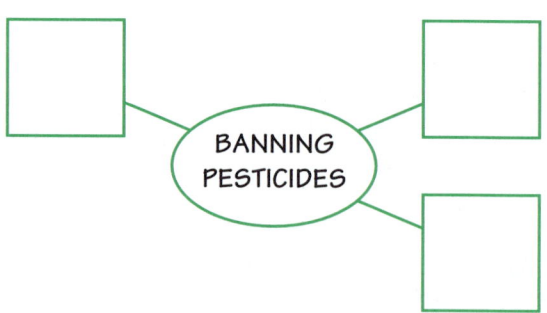

I _____ the Circle of Poison Prevention Act because _____.

Senate Panel Hears of Pesticide Harm Abroad

Costa Rican farm workers blame their sterility on chemicals banned in the U.S. but exported

By Robert L. Jackson
FROM: *THE LOS ANGELES TIMES*
JUNE 6, 1991

Costa Rican farm workers told a [U.S.] Senate committee Wednesday that they became sterile [unable to have children] and suffered emotional trauma [injury or stress] from handling U.S.-produced pesticides that are banned in this country.

The testimony by the workers came as the Senate Agriculture Committee continued a two-year inquiry [investigation] into the alleged [not proven] dumping abroad of U.S. chemicals that have been deemed [considered] too hazardous for use in the United States.

Committee Chairman Patrick J. Leahy (D-Vermont) said his panel is focusing on the pesticide DBCP, which has been banned from domestic use for several years on grounds that it causes sterility. DBCP, also known as 1,2-dibromo-3-chloro propane, is a highly toxic chlorinated pesticide.

California banned DBCP in 1977, and the [Environmental Protection Agency] EPA adopted a nationwide ban two years later, except for use on pineapples in Hawaii. The ban was extended to that crop in 1985.

The Costa Rican farm workers said Wednesday that they never learned the identity of the pesticide they routinely used in the 1970s, but Leahy's staff said that DBCP was widely used at the time in banana fields where the Costa Ricans worked. Its use was discontinued there in 1979, and the chemical is no longer manufactured, the staff reported.

> **"**We have seen contaminated waterways, sick and even dead animals and pesticide-laden foods.**"**

However, Leahy said that "the tale of DBCP is an appalling [shocking] one. In the blind pursuit of corporate profits,

U.S. chemical company giants ignored their own scientists, kept studies secret from their own employees, dumped their poisons overseas and devastated the lives of thousands of unsuspecting and innocent people."

Leahy, who said that some U.S. chemical companies "still export tons of pesticides that the Environmental Protection Agency considers too dangerous for our farmers to use," is sponsoring legislation called the Circle of Poison Prevention Act of 1991, so named because such pesticides used abroad may return to this country on tainted [contaminated] imported fruits and vegetables. It would outlaw the export of pesticides that are banned or unlicensed in the United States.

Testifying with the help of an interpreter, Mario Zumbado, 40, said that he underwent "difficult trauma and severe depression" and saw his marriage break up when he learned after nine years as a farm worker that he could not father any children.

Zumbado displayed a large plunger-type device to show senators how he sprayed the ground beneath the banana plants, sometimes getting spray in his eyes and face. He said that his field supervisors never demonstrated techniques for handling the pesticide safely.

Victoria Sibaja de Zumbado, his second wife, testified that her husband still has emotional problems and depression as a result of his involuntary sterilization.

Waldemar Loaiza, another field worker, said that the pesticides he used often led to a strong, nauseating odor around the small houses where employees lived.

Dr. Roberto A. Chaves, a Costa Rican toxicologist [person who studies the sci-

Senator Patrick Leahy is a strong supporter of the Circle of Poison Act.

ence of poisons], said that the people who worked with DBCP generally are "moody . . . , with poor sex drive and often impotent [unable to produce children]." He added that "among many families with children of working age, the totality of male offspring were sterilized."

Another witness, Dr. Catharina Wesseling, who has worked for eight years in the pesticides program of the National University in Costa Rica, told the panel that "we have found pesticide use in all the Central American countries to be intensive, extensive and thoroughly out of control."

Wesseling added: "We have seen or documented many human poisonings.

We have seen contaminated waterways, sickened and even dead animals and pesticide-laden foods. This is the price Central America is paying for its efforts to become a major agricultural export region and pull itself out of its terrible debt situation."

A spokesman for the National Agricultural Chemicals Association said that the group favors a ban on chemicals that have been forced off the U.S. market but disputes the need for stopping sales abroad of products that only lack a U.S. license. He said that many of the unlicensed products are waiting for approval from the EPA.

4 GETTING INVOLVED

Cooperative Learning

Find Out More Work with three other students and contact the Environmental Protection Agency and ask them to send you information on the harmful effects of pesticides. (The address and phone number can be found in the Community Resource Directory starting on page 244.) In addition, ask them to send information on the process they use to approve pesticides. How do certain pesticides get approved? Make a time line that illustrates the approval process. Be creative!

Gathering Information

Take a Public Opinion Survey As a class, develop a public opinion survey. Include questions such as: Should pesticides banned in the United States be sent to other countries? Should U.S. chemical companies be forced to compensate foreign farm workers for illnesses caused by the use of these pesticides? Try to limit your survey to no more than five questions. Each person in the class should try to get ten different people to complete the survey. After the class has gathered all the responses, summarize the data. Send your results to the Senate Agriculture Committee, in care of the U.S. Senate. (The addresses can be found in the Community Resource Directory starting on page 244.)

Critical Thinking

Investigate Locally Go to your local grocery store and find out what kinds of fruits and vegetables are imported from other countries. Make a list of no more than 15 fruits and vegetables and ask the manager of the produce department if the items on your list are imported from other countries. You also may look at the labels or bags that contain the fruit or vegetable to see if it is imported. How many of the items on your list are imported? Do any come from Central or South America? Is there a chance that any of the imported fruits or vegetables may be contaminated with pesticides? (You should wash all fruit carefully before eating it, regardless if it is imported or grown in the U.S.) Summarize your findings in a written report.

5 FOCUS ON...
Condom Distribution in High Schools

The widespread incidence of AIDS is one of the biggest problems facing society today. AIDS has no geographic or economic boundaries; it affects everyone. More than one-fifth of AIDS victims are under the age of 20. How should the spread of AIDS be dealt with? One method being explored is distributing condoms in high schools. Do you feel this is an appropriate way to fight AIDS? Read the article and see how different groups view the issue.

BEFORE YOU READ

Below is a list of actions that could possibly be taken to prevent the spread of AIDS. Read the actions, then write them on a sheet of paper starting with the action that you think is the most effective and ending with the action you think is least effective. You may include an action of your own as well. In addition, write your reason for ranking the actions the way you did.

Actions

1. Provide information to students about AIDS—what it is and how it is transmitted.
2. Provide free condoms.
3. Provide counseling for students about safe sex practices.
4. Test all students each year for the AIDS virus.
5. Have students sign a pledge agreeing to abstain from sex.
6. Provide counseling for students about self-esteem and sexual abstinence.
7. An action of your own.

WHILE YOU READ

Copy the graphic organizer below onto a separate sheet of paper. As you read the article, keep track of the players who support distributing condoms and those who oppose distributing condoms. Add them to your organizer. Next to each player, write the belief of that player.

Distributing Condoms in High School

PLAYER	BELIEF

Better Safe Than Sorry?

by Susan Tifft, with reporting by Katherine L. Mihok and James Willwerth
FROM: *TIME*
JANUARY 21, 1991

Angry parents, politicians and clergy gathered on the steps of New York's City Hall last week amid placards [posters] that demanded STOP FERNANDEZ FROM TEACHING OUR KIDS GAY SEX and DUMP KING CONDOM FERNANDEZ. What schools chancellor Joseph Fernandez is doing, warned Monsignor John Woolsey of New York's Roman Catholic Archdiocese, amounts to a "ratification of sexual promiscuity [approval of having sexual relations with many people]." Said [an] outraged parent: "Fernandez is insulting our children by telling them they cannot be educated as to what is right."

The indignant [angry] rally was a dress rehearsal for a bigger confrontation this week, when the New York City board of education is scheduled to hold a public hearing on Fernandez's proposal to make condoms freely available in the city's 120 public high schools as part of the battle against AIDS. If the plan is approved, New York City's will become the first school system in the nation to provide condoms on an unrestricted basis—without fees, parental consent or counseling requirements. Fernandez's own family is divided over the issue. "My wife doesn't agree with me," he says.

New York City is the biggest battlefront, but the war over condom distribution in schools is spreading across the country. One high school in Cambridge, Massachusetts, three in Chicago, three in Los Angeles and one in Miami already dispense the devices to students through in-school health clinics, if parents give their consent. Sharon Pratt Dixon, the newly inaugurated mayor of Washington, backed school-based condom programs during her election campaign, provided students receive instruction in human reproduction and safe-sex practices.

> *"This gives a stamp of approval to something we feel is immoral and unhealthy."*

But in most places, the idea has met with anger, outrage—and defeat. Last fall a proposal in rural Talbot County, Maryland, to make condoms available in high schools failed to pass the school board by just one vote. In prosperous Marin County, California, Tamalpais High School abandoned a plan for condom distribution after a coalition of pro-life supporters and parents filed suit to stop it. Los Angeles' pilot reproductive-health project overcame vigorous opposition only when the city agreed to a parental-consent feature; about 75% of parents at the three participating schools have acceded [agreed].

The current debate is the result of a grim trend: the increasing incidence of AIDS among adolescents. Of the 157,525 cases reported to the Centers for Disease Control through November 1990, 615 involve 13- to 19-year olds—154 more than 11 months earlier. That total understates the gravity of the problem. Since more than a fifth of all AIDS victims in the U.S. are in their 20s and the incubation period for the disease can be as long as 10 years, most of the older age group probably became infected as teenagers.

Nowhere is the challenge more grave [serious] than in New York City, which accounts for just 3% of the nation's 13- to 21-year olds, but harbors 20% of all reported AIDS cases in that age group. It was the sheer size of the problem that prompted Fernandez to suggest the free-condom idea as part of an expanded AIDS-education program for the city's 261,000 high school students. Under the plan, staff volunteers at each school would hand out condoms, along with a booklet explaining their use, to every student who wants them. Sex counseling would be available but would not be required, for fear it would deter students from seeking protection.

One of the standard objections raised is that the ready availability of condoms will only encourage teenagers to have sex. "This gives a stamp of approval to something we feel is immoral and unhealthy," says Rabbi Abraham Hecht, president of the Rabbinical Alliance of America. Some parents resent the loss of control over their child's decision; others think the sagging school system could put its dollars to better use. "The chancellor's primary mission is education," insists John Hale, a former member of the New York State board of social welfare. "He's not the health department."

Critics also argue that condoms, which can have a failure rate of between 10% and 15%, are not the best protection against AIDS. . . . The alternative that schools should be promoting, critics argue, is chastity [not having sexual relations].

There is little evidence, however, that sexual abstinence is an attractive option for students. Almost all existing sex- and AIDS-education classes stress chastity, yet half the nation's high school girls are sexually active. . . .

> "This isn't telling us to be sexually active, it's just saying, if you are, you should be protected."

Supporters of the Fernandez plan argue that although condoms are available in drugstores, many teenagers do not use them properly or consistently. That makes it necessary for schools to step in to safeguard the public's health and that of their charges [students]. Parental consent, boosters [supporters] say, is desirable but unrealistic. "I don't know anyone whose children consulted with them before they had sex," says Caesar Previdi, principal of Manhattan's Martin Luther King Jr. High School. At Jordan High School in the Watts section of Los Angeles, teens are trained to counsel one another on sexual issues, precisely because adult advice is so often shunned.

In fact, many parents seem relieved to have the issue taken out of their hands. A Gallup poll for the daily New York *Newsday* found that 54% of parents with children in the New York City public schools approve of the condom plan. There is little opposition from students. . . .

5 GETTING INVOLVED

Cooperative Learning

Set up an AIDS Awareness Fair As a class, set up an AIDS awareness fair. Divide the class into groups. Each group should have a different responsibility. One group may gather information by writing or calling the Centers for Disease Control, the American Red Cross, and the American Foundation for AIDS Research (AmFAR). (Addresses and phone numbers can be found in the Community Resource Directory starting on page 244.) Tell the various organizations what you are doing and request that they send you brochures, posters, and any other information. Another group may call the local chapter of the American Red Cross and request someone to come and speak. A third group of students may want to hang posters advertising the fair throughout the school. For some hints on getting started, see page 243.

Gathering Information

Find Out More Work with three other students to brainstorm three alternatives to preventing AIDS and other sexually transmitted diseases without distributing condoms. Research the alternatives your group has decided on in your school or local library. Provide supporting documentation for each of your alternatives. Present your alternatives to the rest of the class.

Critical Thinking

Discuss the Issue In a group of four students, discuss the following belief statements:
1. The distribution of condoms in school promotes promiscuity.
2. The responsibility for sex education should rest solely on parents or guardians and not on the school system.
3. Schools should be allowed to distribute condoms on an unrestricted basis.
4. Schools should distribute condoms through in-school health clinics if parents or guardians give their consent.
5. If sex education classes stressed chastity (abstinence), the percentage of teenagers who are sexually active would decline.
6. The distribution of condoms in schools would help prevent the spread of AIDS among teenagers.

With which of the beliefs does the group agree? With which of the beliefs does the group disagree? Have one person report the group's consensus (if there is one) to the class.

6 FOCUS ON...
Required Aids Testing

Should every patient and health care worker be regularly tested for AIDS? This is just what is being proposed by Congress. Health care workers feel that mandatory testing is an invasion of privacy and could result in loss of jobs. On the other hand, mandatory testing could allow surgeons to request an HIV test on a patient before the surgeon operates. What happens if patients refuse to be treated by infected doctors? What if doctors refuse to treat infected patients? Do the potential costs of mandatory AIDS testing outweigh the benefits? Will mandatory testing prevent the spread of AIDS? Keep all of these issues in mind as you read the following article and decide for yourself where you stand on this issue.

BEFORE YOU READ

Copy the chart shown onto a separate sheet of paper. Read each of the statements below. Then write them on the chart. Be sure to include the reasons for your beliefs.

In favor of	Opposed to

Statements

1. All health care workers should be tested for AIDS four times a year.
2. Anyone having surgery should be tested for AIDS.
3. Drug addicts should be given clean needles.
4. Beginning in elementary school, children should receive accurate information about the AIDS virus.
5. What other solutions could you suggest?

WHILE YOU READ

Copy the chart shown onto a sheet of paper. As you read the article "HIV Tests in the Health Profession," keep track of the arguments that people have for and against mandatory AIDS testing.

Mandatory AIDS Testing	
Arguments in favor of	Arguments against

HIV Tests in the Health Profession

Some groups say mandatory screening's benefits wouldn't justify cost

by Malcolm Gladwell
FROM: *THE WASHINGTON POST*
SEPTEMBER 11, 1991

Every year millions of Americans are prodded, probed or cut open in dental offices and hospitals in such a way that their blood could come in contact with the blood of an attending physician, nurse or dentist.

Because the AIDS virus can be transmitted under these circumstances, should every patient and health care worker be regularly tested for HIV?

In July an amendment offered by Senator Jesse Helms (Republican-North Carolina) effectively requiring such widespread HIV testing overwhelmingly passed the Senate. But in the past few weeks, as the House [of Representatives] and Senate prepare to confer [meet] over the Helms amendment, some health care groups and AIDS activists have begun to argue that the potential costs of widespread HIV testing far outstrip the benefits.

Just how expensive health care testing would be is unclear, dependent on the assumptions used. But estimates generated by the Service Employees International Union, the AIDS Action Council and researchers at San Francisco General Hospital suggest that the Helms bill, if enacted, could add at least another $1 billion to the annual U.S. health care tab while saving, at best, a handful of lives.

"No matter how you look at it, this is going to require an incredible amount of human and financial resources," said Julie Gerberding, the chief author of the San Francisco study. "If we assign this proportion of the pot to address this problem, there is going to be much less money available to deal with some of our other more important health problems."

At issue are two amendments proposed by Helms just before the congressional recess. The first would impose criminal sanctions [charges] on any HIV-infected health care worker who performed a procedure on a patient where there was a possibility of blood-to-blood contact without first informing the patient. Although the measure would not explicitly [state clearly] require HIV testing, the assumption of many health care experts is that liability questions [who is responsible] created by the bill will force hospitals to require regular testing of their employees.

The second amendment would allow any surgeon about to perform an invasive procedure to request an HIV test of the patient without securing patient permission.

No attempt was made when the bills were drawn up to calculate the precise economic impact of the measures. But proponents [supporters] of broader testing argue that it is a prudent [wise] step to minimize the spread of HIV infection in the health care setting.

Mike Franc, an aide to Representative William E. Dannemeyer (R-California), who has proposed similar if less sweeping legislation in the House, said that hospitals already test quite frequently for HIV so that extra testing would not add that much of a burden. In addition, he pointed out that if done on a large scale, the costs of HIV testing would fall substantially.

"The fundamental question is not what does it cost to test, but what has it cost not to test," said Shepherd Smith, president of Americans For a Sound AIDS Policy. "We have an epidemic out of control and I think in large part it is because we haven't tested in the past."

Others, however, are less sure that a widespread testing program makes as much sense as simply adhering to current government regulations requiring that "universal precautions"—like rubber gloves, sterilization and safe operating techniques—be used to prevent the transmission of the virus. Government health care experts also believe there is little evidence that adding mandatory testing to universal precautions will help prevent the spread of infectious disease.

Of the approximately 6.6 million health care workers in the United States, analysts at the AIDS Action Council estimate that about 1.1 million come into close enough contact with patients to make them candidates for regular HIV testing. If they were all tested for the virus four times a year—as some have recommended would be necessary—that could come to $220 million a year.

The Service Employees Union, using different cost assumptions and assuming that the number of health care workers being tested would be closer to 4 million, came up with an estimate of as much as $1.5 billion.

The cost of testing patients is potentially higher. If all 33 million Americans admitted to a hospital every year are tested, the total bill could be $1.65 billion. If just those about to undergo an operation are tested, the bill could be several hundred million dollars.

"*We have an epidemic out of control.*"

All of these estimates assume that the cost of providing an HIV test is approximately $50. This figure, however, includes the cost of pre- and post-test counseling, which is routinely given both to people who test positive and negative for HIV. . . .

In her paper, however, which was published this summer in the journal *Infection Control and Hospital Epidemiology*, Gerberding points out that even if testing and counseling costs can be reduced over time, there are many other burdens that would follow from widespread testing.

Hospitals would have to monitor and keep track of test results. Infected workers, if they could no longer be permitted to do their jobs, would have to be replaced, reassigned or retrained. Identifying people as HIV positive who might otherwise have kept their status private makes them formally eligible for worker's compensation. Former patients of infected employees would have to be notified and tested.

Adding all these expenses, Gerberding came to an average cost of somewhere around $1 million per year for an average-sized urban hospital, assuming the hospital tests four times a year and uncovers at least one infected employee per year. Extrapolated [estimated] nationally, she said, this cost would be "staggering"—topping $1 billion a year.

The $1 million estimate, she noted, is more than double what a typical hospital currently spends on controlling all hospital-based infections.

6 GETTING INVOLVED

Decision Making

Write a Letter to Your Senator As a class, decide on a group consensus of the two amendments proposed by Jesse Helms. Write a letter to one of your senators asking his or her opinion on the Helms' amendment. State the class's opinion on the two amendments in the letter. (Addresses for the U.S. senators can be found in the Community Resource Directory.) For some guidelines on writing to a member of Congress, see page 242.

Cooperative Learning

What Would They Say? Get together with three other students. On index cards, create a statement from each of the six individuals below. Each quotation should have enough of a clue within it so that another group can identify the speaker from those listed below. Exchange your cards with another group of students. Be sure to keep track of your group's answer key so that you can check the challenge group's work. Number each response and key it to the number of the individual below.

1. AIDS Action Council president
2. HIV-infected health care worker
3. HIV-infected patient
4. health care worker
5. hospital patient about to be operated on
6. hospital administrator

Critical Thinking

Draw a Conclusion Work with another student to make a list of the benefits and costs of mandatory AIDS testing. Develop a numerical scale with numbers from 0 to +5 as benefits of mandatory testing, and from −5 to 0 as costs of mandatory testing. Use the scale to assign a number value to each of the benefits and costs on your list. Add up all the numbers until you get a final score. What is your final score? What conclusions can you draw from your results? Write a brief summary of your conclusions.

7 FOCUS ON...
Animal Testing

For over 1,000 years, scientists have used animals for medical research and testing. Animal research has provided treatments and cures for many once-fatal diseases. Almost every person who has received an injection or immunization for a disease has benefitted from animal research. However, animals are also used for unnecessary testing, such as finding out how irritating a certain eye-shadow may be. The following two editorial articles represent the two different sides of this issue. One is written by a doctor and the other by a member of an animal rights group. Read these articles and decide how you feel on this issue.

BEFORE YOU READ

Below are some possible sources that you could use to learn about animal testing. On a sheet of paper, write the three sources you would most likely use to decide whether or not animals should be used for medical testing. Next to the source, explain your reasons for picking it. In addition, add one other possible source that you would use.

Possible Sources of Information

1. newspaper articles about current testing that is being conducted using animals
2. interview of a doctor about medicines that the doctor finds useful and were developed through animal testing
3. a magazine published by PETA (People for the Ethical Treatment of Animals)
4. pamphlets prepared by the government that describe its guidelines for the use of animals in medical testing
5. interview of a person passing out literature that opposed animal testing
6. your own idea

WHILE YOU READ

Copy the chart shown below onto a separate sheet of paper. Use the chart to help you analyze the arguments for and against animal testing. Fill in the chart as you read the following two articles. After you have read both articles, put an asterisk (*) on the side that you feel has the stronger argument.

For Animal Testing	Against Animal Testing

Behind Animal Testing Is Profit Motive

By Valerie W. Gaston, member of People for the Ethical Treatment of Animals
FROM: *ROANOKE TIMES & WORLD-NEWS* OCTOBER 18, 1991

. . . It does not take a Ph.D. to read the many articles, studies, books, etc., written by veterinarians, researchers and other experts in related fields to understand that much animal research is barbaric [uncivilized], sadistic [causing pain] and totally useless and that there are many options available—which brings us to the main reason these options are not utilized: greed. Animal testing is much more profitable to companies, plus many "researchers" would lose their government grants and corporate funding were they to cease their practices.

The typical animal used for research is kept in a small cage, with little or no food and water. They [the animals] often live in their own feces and urine, which they often consume due to hunger and thirst. They are deprived of social interaction and exercise.

Many dogs and cats used were once family pets that were stolen or lost and ended up in shelters, etc. All of these physical and psychological factors affect the animal's behavior and physiology, making the validity of test results questionable.

Test results have proven that other methods of product testing are as accurate if not more accurate than using animal models. Computers, for example, can be utilized to accurately give the toxicity of substances [how harmful they are]. They can be used to determine the actual structural makeup of chemicals. . . .

This program can replace animals when performing LD50 and Draize eye and skin tests. It can also be used to predict carcinogenicity [kahr-si-nu-je-NIS-u-tee] and mutagenicity [myoot-u-je-NIS-u-tee, whether the products are cancer-causing or will cause genes to change]. Other methods available are simulated tissues and body fluids and live cell cultures. The use of cadavers [dead bodies] has been very successful.

All of the above methods have proven reliable, yet literally millions of animals have been/will be burnt, electrocuted, beaten, starved, tormented, etc., in the name of science this year alone. It is totally beyond comprehension.

For years this country has boasted about our technological advances—but injecting a rabbit with toxic cleaning fluid, which can cause seizures, paralysis and a very slow, painful death, or dropping a live rat into boiling water, or crushing a dog's hind legs to measure how long it will take until the animal goes into shock or dies from the pain, does not sound very advanced to me; it is sick and barbaric.

It is long past time for the public to be educated regarding animal testing. Millions of animals have been tormented and have suffered excruciatingly [unbearably] painful deaths. This must be stopped. If it is not, the next time a pet of yours disappears, it too could be facing a fate worse than just death in a research lab.

Think Animal Testing Is Inhumane? Explain It to the Sick, Dying

by Dr. Kurt J. Isselbacher,
Harvard Medical School Professor and Director of Massachusetts General Cancer Center
FROM: *CHARLOTTE OBSERVER*
JUNE 12, 1991

Anyone who says experiments on laboratory animals are unnecessary should explain how physicians learned to treat President Bush's recent illness. The radioactive iodine he took so physicians could scan his thyroid gland was developed through research on rats and larger mammals. The anti-coagulation drug he received to prevent blood clots has been tested on rabbits, swine [pigs] and other lab animals. Many other aspects of his treatment for Graves' disease also were developed with animal research.

The same is true of current studies of thyroid and heart problems. A research team at University of North Carolina Medical School is using rats to learn about hypothyroidism [thyroid disease characterized by slow metabolism]. Animal studies at Columbia University, Case Western Reserve University, Washington University and elsewhere may help physicians learn about atrial fibrillation, the irregular heartbeat President Bush experienced.

The president's case is notable but not exceptional. At the age of 3, a much less publicized patient named Charlotte Evert had potential fatal narrowing of the arteries. She underwent a new procedure that widened her arteries and improved her blood flow. The procedure, developed by a physician using dogs and human cadavers [dead bodies], gave Charlotte a normal life.

Greg Maas, a father of two, underwent chemotherapy to overcome a form of cancer that once was invariably fatal. The drugs he took were tested on mice.

"TREADMILLS! MAZES! THERE MUST BE MORE TO LIFE THAN THIS."

Everyone Benefits

There are millions of examples like these. Anyone reading this article has benefitted from animal research that led to vaccines against deadly diseases, treatment for infections and virtually every other medical advance in this century.

These stories need to be told because advocates of "animal rights" are threatening the efforts of scientists to develop better treatments not only for thyroid and heart problems, but for cancer, Alzheimer's disease, AIDS and other afflictions.

As a committee of scientists that I chaired for the National Academy of Sciences and Institute of Medicine concluded recently, animal research remains an irreplaceable cornerstone of efforts to improve human health. Abandoning it would deny new medicines and cures to future generations.

If human beings had chosen a century ago to stop using animals in research, the world would be a different place today. Many of us are alive because diseases were controlled through the knowledge gained from animal research.

This research remains essential, and it is much more limited than its critics portray. The number of vertebrate animals used each year in research, education and testing is a fraction of 1 percent of the number killed for food. Eighty-five percent of the animals are rats and mice. Comparatively few dogs and cats are used, and they come mainly from animal shelters and pounds—which have so many unwanted dogs and cats that they kill approximately 100 for every one provided to scientists.

Alternatives are Welcome

Researchers would welcome the opportunity to use tissue cultures, microorganisms, computer models and other alternatives in place of animals, which are expensive and inconvenient. Replacements for animals have been developed for some kinds of experiments, and the search for alternatives is continuing. But, for about half of the biomedical investigations carried out in the United States, animals remain essential.

Researchers have an obligation to minimize the pain and distress of lab animals, and to see they are used only for productive goals. On the rare occasions when they violate this trust, they should be disciplined. Indeed, "animal welfare" proponents [supporters] have performed a valuable role in helping ensure that research animals are treated humanely.

But animal rights advocates, who go much further and argue that scientists should abandon these experiments entirely, owe an explanation to terminally ill children and millions of other Americans waiting for biomedical advances. They should go to their bedsides and tell these patients why they are less important than the animals.

7 GETTING INVOLVED

Cooperative Learning

Investigate Further In groups of four students, visit your local grocery or discount department store. Find the cosmetics section. Examine the labels and write down the names and addresses of five companies that sell cosmetics. Look for toll-free phone numbers, too. Write a letter or call each cosmetic company and ask for information concerning its product-testing procedure. Ask about the procedure for a few different types of products. For example eyeshadow, mascara, face cream, sensitive-skin makeup, and so on. Is the product testing done on animals? Compare the responses from the different companies. Summarize your findings on a poster.

Decision Making

Have a Forum Invite each of the following people to be a guest speaker in your class: a medical doctor, a veterinarian, an employee of the ASPCA, and an animal rights activist. Decide what kind of information relating to animal testing you want to know, and decide on appropriate questions for each person. Send a list of questions to each person before he or she comes to class so that he or she can prepare answers.

If you cannot arrange for speakers to come to your class, write a letter to each person requesting the same information. For some hints on writing a letter requesting information, see page 234. After all your questions have been answered, determine an action the class may want to take on this issue of whether animal testing should be allowed.

Critical Thinking

Separate Fact from Emotion Both of the articles were "letters to the editor." The purpose of a letter to the editor is to persuade, or influence, how the editor and readers think about a particular issue. Sometimes letters to the editor contain emotional descriptions as well as facts. Read through both letters again. This time, try to separate the fact from the emotion-producing statements. Make a list of the facts contained in the letters. Once you have separated the facts, try to write an article on the issue containing only facts.

8 FOCUS ON...
Genetic Engineering and Cystic Fibrosis

Cystic fibrosis (CF) is an inherited disease characterized by an abnormal build-up of thick liquid in the lungs, the pancreas, and other organs. A person with cystic fibrosis usually lives only until the age of about 28. With improved **genetic engineering** techniques, the gene that carries cystic fibrosis was discovered. Genetic engineering is a form of genetics in which scientists work directly with individual genes. Scientists are now looking for new methods of altering or replacing the defective gene. Read this article to find out more about these new discoveries.

BEFORE YOU READ

Work with another student. Using what you already may know about cystic fibrosis, predict possible endings to the following phrases. Then after you have read the article, see how closely your predictions matched the article.

1. For half a century, doctors have been testing _____.
2. One drug promises to prevent _____.
3. In the past two years alone, researchers _____.
4. Another therapy may be coming along sooner _____.
5. The discovery of the CF gene has revolutionized _____.

WHILE YOU READ

As you read the following article, you will learn about different types of treatments for cystic fibrosis. Copy the chart shown onto a separate sheet of paper. Fill in the chart to help you keep track of the various treatments. Compare your chart with that of another student. Do the charts agree?

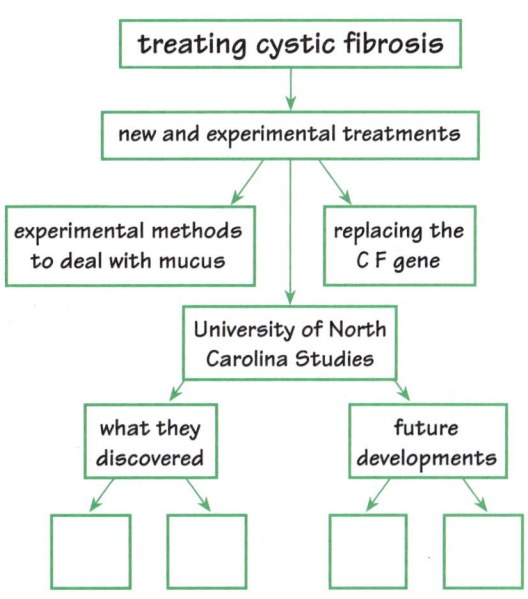

169

Laying Siege to a Deadly Gene

By Andrew Purvis
FROM: *TIME*
FEBRUARY 24, 1992

... Cystic fibrosis [is] the most common inherited disorder among whites and a disease that afflicts 25,000 Americans, killing more than 500 every year. Just 10 years ago, the prospects [for survival] were as bleak as they were inevitable ...

In the past two years alone, researchers have reported preliminary success with two separate therapies that for the first time treat the underlying cellular disorder as opposed to just the symptoms of the disease. More promising still, doctors are closing in on a technique for replacing the defective CF gene, which was discovered in 1989. The discovery has spawned [given rise to] an unprecedented [unheard of] proposal to screen tens of millions of Americans for the defect, so that couples can avoid having an affected child. ...

For half a century, doctors have been treating cystic fibrosis symptom by symptom, doing their best to stem the rising tide of mucus triggered by the disease. As this abnormally thick fluid builds up in the lungs, pancreas, liver and other organs, it not only serves as fertile ground for damaging infections but also blocks the passage of vital digestive enzymes [protein that controls chemical reactions in the body] to the intestine and stops up sperm in the testes. As a result, patients have difficulty breathing, digesting food and even reproducing. ...

Using simple therapies, such as clapping young patients on the back and chest several times a day to clear the lungs or providing a special nutrient-rich diet, scientists have made impressive strides against the ailment, extending the average life-span from just five years in the 1950s to 28 today. Recent developments in genetic engineering may refine this tactical approach still further. One drug promises to prevent a protein called elastase, produced in dangerous quantities by the CF patient's own immune cells, from attacking lung tissue. Another synthetic enzyme called DNase instantly dissolves

---

A DECADE OF DISCOVERY
In the past decade, the understanding of cystic fibrosis has progressed from a symptom-by-symptom view of the disease to a grasp of its cellular and finally its genetic roots.

Adapted from: *Time*

Before the 1980s: Doctors understood and treated only the symptoms, including the accumulation of mucus in the lungs and pancreas, chronic infections and poor digestion.
Typical Treatments: Clapping on the back, chest and sides to loosen phlegm; special diets; supplemental enzymes; antibiotics.

leftover DNA from dead immune cells, one of the bulkiest components of the accumulating mucus. The thinner fluid can then be cleared by the body's own self-cleansing mechanisms.

But even such sophisticated mucus busters would do little to stop the buildup of fluid at its source. In the early 1980s, scientists at the University of North Carolina opened the way for a radically new kind of therapy that would attempt just that. The doctors noticed that cells taken from the lungs of CF patients contain abnormally high levels of sodium and chloride—the constituents of [elements that make up] salt. This did not come as a complete surprise, since CF patients' sweat is known to be abnormally salty, a sign that their bodies do not handle the mineral properly. But the U.N.C. researchers realize that this imbalance in the lungs could explain why thick mucus was accumulating [building up] there. An excess of salt within cells was leaching water out of the mucus. [The level of salt in the cells is so high that it absorbs water from the mucus.] This apparently was the basic defect behind the disease.

In the past 24 months, the U.N.C. team hit upon two drugs that could help repair this cellular malfunction. One, a blood-pressure medication called amiloride, slows the uptake of sodium. The other, containing naturally occurring substances called ATP and UTP (for adenosine and uridine triphosphate), stimulates the secretion of chloride. Both have proved effective in early trials, although a marketable treatment is still several years away. . . .

Another therapy may be coming along sooner than [Michael] Knowles [a CF researcher at U.N.C] thinks. Since Francis Collins and Lap-Chee Tsui discovered the CF gene on chromosome 7 in the summer of 1989, researchers from around the world have been struggling to devise a way to bring that finding to the bedside. The challenge: to transport corrected versions of the DNA into the lungs of CF patients. Dr. Ronald Crystal at the National Heart, Lung, and Blood Institute believes the best vehicles are ordinary cold viruses, which (as most people know too well) have a special fondness for the linings of the airways. Ordinarily, these viruses infect host cells by injecting their own DNA through the targets' outer membranes. Crystal hopes to harness this propensity by first disabling the microbes, so that they no longer cause colds, and then inserting a corrected version of the CF gene into the viral DNA.

. . . The discovery of the CF gene has revolutionized the diagnosis of CF. Some public health experts believe that since doctors can identify the defective DNA (which occurs in 1 out of 25 Americans) they should screen all prospective parents. Men and women who find that

Early 1980s: Scientists discovered that cells lining the lungs are high in salt, drawing water from the surrounding mucus. This explains the buildup of thick mucus.
Experimental Treatments: ATP/UTP and amiloride, which help restore the balance of salt in these cells.

1989: Gene is discovered.
Future Treatments: A genetically engineered cold virus containing a replacement for the defective gene could be sprayed into the lungs.

NOW: Capitalizing on the gene discovery, researchers have identified the protein in the cell wall that causes the salt imbalance and are working to produce quantities of it.
Future Treatments: An aerosol spray that provides patients with the protein they lack. Drugs that enable the defective protein to work properly.

they are both carriers might then choose to adopt or conceive with donor sperm or eggs. Last November [1991] the National Institutes of Health financed a handful of pilot projects to help it decide whether a massive screening program would be worth the considerable cost.

Many doctors are not so sure. "Just because we technically know how to test for the DNA, doesn't mean we are ready to do this on a large scale," asserts Collins. The test is imperfect, he notes; it picks up just 85% of carriers. A positive result, moreover, means only that the couple has a 1 in 4 chance of having a baby with the disease. Without proper counseling, Collins says, people might feel needlessly alarmed. . . .

8 GETTING INVOLVED

Cooperative Learning

Debate the Issue Work with four other students to debate the issue of mass genetic screening. One person should be the moderator, two students should take the stand against genetic screening, and two students should take the stand for genetic screening. Using the facts found in the article, as well as your own opinion, prepare a one-minute statement concerning mass genetic screening. Develop at least three questions to ask the opposing group. After each opening statement has been given, continue to debate the issue. After ten minutes, the moderator should decide which side had the most convincing argument.

Gathering Information

Find Out More Call your local chapter of the Cystic Fibrosis Foundation and request a speaker to come to your class to speak about cystic fibrosis and any new methods of treatment. (The phone number and address for the Cystic Fibrosis Foundation can be found in the white pages of the telephone directory.) If a speaker is not available to come to your class, request that some informational pamphlets and booklets be sent to your class. Develop a bulletin board or other type of display with the information you receive. Write a summary of what you have found out.

Critical Thinking

What Do You Think? Answer the following questions in complete sentences. Remember, there are no right or wrong answers—these are your belief statements.
1. Should animals be used in genetic tests that may lead to the prevention or cure of human diseases?
2. Should gene manipulation (altering or replacement of the natural genetic code) be an accepted medical practice? Why or why not?
3. Should married couples be required to undergo genetic testing and genetic counseling before having a child? Why or why not?

9 FOCUS ON...
Medical Testing and Women

In the past, the medical profession did testing and research based on what they felt a group's "contribution to society" was. As a result, all testing was done on white, middle-class, middle-aged men. Tests were never performed on women, minorities, or the aged. If a woman was found to have a disease, such as heart disease, physicians would give her the same treatment and medication that was found to be effective in a man. Today, women legislators and scientists realize that there is an inequity in funding for testing and are trying to do something about it. Should one person's "contribution to society" be a basis for research and money allocation?

BEFORE YOU READ

Copy the chart shown onto a sheet of paper. Examine each of the words below. In the column headed "medical research," list the words most closely associated with medical research. Do the same for the other two columns. Use each word only once. When you have completed the chart, see if you can predict what the article might be about.

Words

heart disease	medication
diagnostic testing	policy making
breast cancer	research grants
osteoporosis	discrimination
National Institutes of Health	
Women's Health Equity Act	

Medical Research	Women's Diseases/ Concerns	Lawmakers

WHILE YOU READ

Copy the chart below onto a separate sheet of paper. As you read this article, pay attention to the reasons why many people feel a new women's health bill is necessary. What are the concerns of women's health research and money distribution before the presentation of the Women's Health Equity Act of 1991? How will this bill benefit research and funding for women's health? What would be the long-term benefits or outcomes of the necessary changes in women's health research?

Women's Health Equity Act

- Why Changes are Necessary (Before the Bill)
  - Research
    - _____
    - _____
    - _____
    - _____
  - Funding
    - _____
    - _____
    - _____
    - _____
- Benefits of the Bill
  - Research
    - _____
    - _____
    - _____
    - _____
  - Funding
    - _____
    - _____
    - _____
    - _____

Group, Legislations Target Women's Health Research

By Michael Clements
FROM: *THE DETROIT NEWS*
APRIL 8, 1991

A group of women lawmakers, scientists and health officials and some of their male allies are taking on the male-dominated world of medical research, demanding more attention to women's concerns.

Women routinely have been left out of major studies of heart disease and tests of drugs. The National Institutes of Health [NIH], the major source of federal money for medical research, spends only about 13 percent of its $7.6 billion budget on women's health. Illnesses that are considered "women's diseases" are often at the bottom of the list for research dollars.

Representative Mary Rose Oakar, D-Ohio, charges that the record represents "a form of discrimination against the whole subject of women's health."

Oakar is a member of a small contingent of women lawmakers backing the Women's Health Equity Act of 1991. The sweeping women's health bill is made up of 22 proposals that would require women to be included in federally funded studies of illnesses and drugs. It also would set up research into breast cancer, ovarian cancer, osteoporosis and other diseases affecting women. . . .

"What you find is not only the research itself not done on women, you find discrimination against researchers who happen to be female and you find that in policy-making there are very few women," Oakar said.

Representative Patricia Schroeder, D-Colorado, said the medical establishment has "failed women across the board. It's failed them by not including women in the research. And it's failed them in not delivering things that if it had affected a majority of the Congress would have been included in normal health-care coverage." . . .

For years, women have been excluded from many major studies of diseases and unapproved medications. A 1988 study on the value of taking aspirin to prevent heart disease, for instance, was based on a study of 22,071 men and zero women.

But heart disease is increasing among women and is the leading cause of death for American women.

Until recently, however, it has been difficult to win research grants to study how heart disease affects women. . . .

Critics also charge that diseases affecting women primarily, or exclusively have been given comparatively little attention.

Osteoporosis, which involves excessive loss of bone tissue, is cited as a dramatic example. The condition affects half of all women over age 45, and the incidence rises to 90 percent among women over age 75. . . .

Mary Fran Sowers, an assistant professor of epidemiology [study of the cause and treatment of different diseases] at the University of Michigan, suspects the policy-makers' interest is relatively low because "it's a disease of women and it's a disease whose ultimate manifestations [effects] are observed in women who are elderly and who no longer are viable contributors to the workforce."

Yet in the aging American population, the impact of osteoporosis "is going to be unbelievable," said Sowers, who is conducting research on the disease.

At the National Cancer Institute, $7.9 million was allocated in 1990 for research into ovarian cancer—an estimate critics also cite as disproportionately low.

"Ovarian cancer ends up being a terminal [deadly] disease for women because there is no diagnostic test for it," said Representative Olympia Snowe, R-Maine, a sponsor of the Women's Health Equity Act. "By the time it is diagnosed, it's generally in the last stages and a woman ends up dying."

The women's health legislation would authorize $30 million a year for basic research during 1992–96 to develop an early detection test for cancer of the ovaries and to determine if the disease has a genetic basis.

9 GETTING INVOLVED

Gathering Information

Library Research Go to your school or local library and research one of the diseases listed in this article. Find out about the causes of the disease, what, if anything, can be done to treat the disease, and how the disease affects the body. Write a brief summary of each of the two articles you choose.

Critical Thinking

Make the S-T-S Connection Think about the article you have just read. Express as clearly as you can the S-T-S interactions. Organize your ideas into a diagram, using arrows to show the flow of ideas between scientists, technology, and members of society.

Decision Making

Determine the Criterion Get together with three other students. Imagine your group is in charge of advising the NIH where to allocate money for medical testing and research. What diseases should be researched further? What groups of people should be included in which tests? As a group, decide on five diseases that should be researched. For each disease, write the groups of people who should be involved in the testing. Compare your list with the lists of the other groups in the class. Which items did you agree on? Which did you differ on?

Human Health and Disease

THINKING CRITICALLY ABOUT LOCAL ISSUES

You may want to work with a group of students for this activity. If so, make sure that each member of the group agrees about the issue that is chosen.

Prioritize the Issues Make a list of health issues that are important to you and your community. Get started by reviewing the brainstorming list that you made at the beginning of this module. If you made a scrapbook of health-related advertisements, look over those as well. If you wrote in a journal about health in developing countries, you may want to include those issues also. In addition, review the magazine and newspaper articles in this module. Evaluate the issues on your list. Then, put them in order so the issue that has the highest priority for you is at the top of the list.

Make the Issue Your Own Become an expert on your issue. Then decide what you as a citizen can do about the issue. Apply the process that you learned in the *Problem Solving* module in the front of this book. Just as a reminder, the skills in the process are:

- Analyzing the Issue
- Gathering Information
- Making a Decision
- Planning Action

If you are working in a group, each group member has to contribute at all four stages. After completing the process, share your experience with your class.

CREATIVE THINKING

Imagine that the National Institute on Drug Abuse has hired you to design a poster. The poster will be used to heighten awareness among school-age children of the dangers of drug abuse. What types of information should you include? Make the poster colorful and informative.

Wrap-Up

HELP WANTED

Research a Career Select a health-related career that interests you. It may be one that was mentioned in any of the articles from this module or one that you already know about. Some examples are druggist, physical fitness instructor, medical doctor, physical therapist, and medical technologist. Collect as much information as you can about the career. If possible, also interview a person in the career to obtain additional information. Then, organize all of the information, including any charts, graphs, or other information, in the form of a poster. Present your career poster to the class. Invite the class to ask questions about the career you have chosen and see if you can answer their questions.

"Just Say No" clubs are groups of students who work to make their schools and communities drug-free. Events like the annual "Just Say No" Walk Against Drugs help create an environment that supports their commitment.

WORLD
Food Resources

> "We live in a highly industrialized, urban culture, but it is important to remember that there is no such thing as a 'post-agricultural' society."
> —Timothy Weiskel, Harvard Ecological Anthropologist

WORLD Food Resources

IN THIS MODULE

Overview	181
Discovery	188
1 "Famine in Africa, the Other Desert Crisis" from *The Arizona Republic*	190
"World Hunger Is Persistent But Not Inevitable, Says New Report" from *The Christian Science Monitor*	192
"In a Changing World, Little Has Changed for the Hungry" from *The St. Petersburg Times*	195
2 "Hunger Said to Afflict 1 in 8 American Children" from *The Washington Post*	197
"Survey: 160,000 Georgia Children Go Hungry" from *Atlanta Journal*	200
3 "Sustainable Agriculture More Pragmatic Than Organic Farming" from *St. Paul Pioneer Press*	203
"Organic Farming Is the Solution to Residue Pesticides, Expert Says" from *St. Paul Pioneer Press*	205
4 "State's Growth Threatens Way of Life in Rice Towns" from *The Los Angeles Times*	207
5 "Irradiated Food Coming, But Not Without Protest" from *The New York Times*	213
6 "Genetic Diversity Prevents Blight, Spread of Famine" from *The Star Ledger*—Newark, New Jersey	220
7 "Scientists Use Gene Alterations to Make Crops Resistant to Infestation" from *The Los Angeles Times*	224
Wrap-Up	228

WORLD FOOD RESOURCES

What kind of peanut butter do you like best—smooth or crunchy? Which brand is your favorite? Most supermarkets probably have at least five different brands, each in two textures in at least three different-sized containers. In the United States and other developed nations, many people take it for granted that they can buy exactly the kind of food they want. Variety doesn't stop with peanut butter. Walking through the produce section of a supermarket, a person can find strawberries, mushrooms, oranges, and melons any time of year. From one aisle to the next, the store is bursting with a variety of foods. People can choose exactly the types of cereals, snacks, and drinks they like best. It can even be taken for granted that all of this food is clean, fresh, and wholesome.

People can not only pick which types of foods they like best, they also pick the store. If the closest supermarket or grocery store is not clean or efficient, or may be too expensive, a person need only go to the next one. Competition among different stores can help keep prices lower.

In addition, all these local stores stock items from around the world. The coffee may come from Colombia and the

In the United States, we are lucky to have a large variety of many items, including produce.

181

olives may have been grown and processed in Spain. Even fresh foods may come from a great distance. A store may stock cheese from Switzerland, tomatoes from Israel, and grapes from Chile. Shipping food from around the world ensures a selection of the best products from each region of the world. But worldwide shipping requires the use of energy, usually **fossil fuels,** the supply of which is limited and expensive. Products that can be grown locally, such as some fruits and vegetables, can be bought from nearby farms. This saves both energy and money.

To buy any of the products grown near home or from other areas of the world, all that is needed is money. All parts of the United States have abundant food available. Yet in many parts of this rich country, people go hungry. They cannot afford to buy enough to eat. Food stamps or other programs funded by the federal and state government, to assist persons in need may not be sufficient to adequately feed a family.

The Source of Your Food

Most of the food—fresh, canned, and frozen—in local stores comes from highly mechanized farms in the United States. Here fields are plowed, and harvested using large, modern machinery. Chemicals to kill weeds and insects and to fertilize are routinely applied. Fields may also be irrigated to ensure enough moisture for the growing crops. This type of farming results in large **crop yields,** that is, large amounts of crops per acre.

Without using machines, chemicals, and irrigation, crop yields would depend almost completely on natural forces. A lack of rain or an infestation of insects could completely wipe out a crop. But the highly technological farming industry of the United States produces so much food that there is an oversupply. Mountains of cheese are stored in caves. Grain is kept for years in warehouses called grain elevators. This excess food can be used when less food is harvested and can be sold to other countries. The problem, of course, is in getting the food to all the people who need it. Various social, political, and economic forces can prevent this from happening.

Although farming in the United States and other industrialized countries is very successful, there are problems. One problem arises from the chemicals used in farming. **Pesticides** and **herbicides,** used to poison insects and weeds, do not stop poisoning once they have been used. They are taken in by wild plants and animals, and wash down into groundwater, into rivers, and out to sea. Some amount is taken up in the food crops themselves and is then eaten by humans.

Another problem is the uncertainty of the availability of water. In the United States, cities are now competing with farmers for water resources. Heavily irrigated lands often become unusable when irrigation waters dry up, leaving salty deposits behind on the land. In areas such as the Middle East, countries argue over the water rights to rivers that flow through more than one country. In the United States, different states also argue over water rights to rivers that flow between them or through one state to another.

A third problem comes from the fact that through technology we have increased the yields of a relatively small number of crops. For economic reasons, farmers have concentrated on planting these high-yield crops. High-yield crops

Many farmers grow wheat because it is a high-yield crop.

bring in more money. The resulting lack of **gene diversity** could cause entire harvests to be destroyed by disease or adverse weather conditions.

When farmers plant only one type of crop, they need to plant large areas at the same time. It is not until these seeds sprout and the plants develop sturdy roots that there is a guard against soil loss. Soil can be carried away by wind or water. If a very heavy rain falls on a freshly plowed field, large amounts of valuable topsoil will be washed into nearby streams. Not only is the land left without valuable nutrients, but the streams are clogged with silt and possibly polluted with pesticides, herbicides, and fertilizers. **Deforestation,** or the elimination of many trees in an area, has resulted in floods that have damaged harvests.

Subsistence Farming

Many parts of the world are not faced with the problems of industrialized agriculture. Instead of an overuse of chemicals and fossil fuels, there is an undersupply of farming aids. In **subsistence agriculture,** energy is provided by the sun, humans, and farm animals. Four billion people depend on subsistence agriculture. They grow just enough food for their own survival.

Only a little over 2% of the people in the United States are farmers. Yet, these people produce more than enough food for the remaining 98% of the population. In contrast, subsistence agriculture involves the whole society. Instead of going to a well-stocked and brightly lit supermarket, each family raises its own food. In good times, the family produces enough food. When possible, extra food

While only 2% of the people in the United States are farmers, many people grow their own vegetables.

is traded or saved. But when conditions are not good, too little food is produced.

An extreme shortage of food over a period of time is called a **famine.** Famines can be the result of natural disasters, such as floods, droughts, plagues, pests, or diseases that kill crops or livestock. Famines often occur in developing countries, where subsistence agriculture is common. Families' small reserves of food are not enough to last until the next harvest. In the case of droughts, it can be several years between crops. So developed countries with large supplies of excess food sell or donate food to the countries in need. For example, the former Soviet Union bought supplies of wheat for a number of years. Problems arise when the food does not arrive quickly enough. Also, long-range problems are not solved by the short-term answer of donating food. Without the technology that would enable these countries to produce enough food for themselves, famines will continue to occur.

Famine can also be caused by civil wars. Unfortunately, food, or rather the lack of food, is often used as a weapon. One way to win a war is to starve the opposing soldiers as well as civilians. In this case, there is little chance of donated food reaching those in need. Wars can also cause the destruction of usable cropland and deforestation, which can lead to soil erosion.

Food Problems

Although there is excess food in some parts of the world, it is not available to all who need it. The developing countries are home to 80% of the world's population but produce less than half of the world's food supply. As a result, these people generally get less food.

Food provides energy. The energy in food is measured in **calories.** You probably have seen calorie counts on food packages. In the United States, many people watch their calorie intake. Too much food and too little exercise can lead to many different health problems.

However, eating too little is an even greater problem. People can suffer from undernutrition—taking in too few calories. Others suffer from **malnutrition**—taking in enough calories, but not enough nutrients needed for good health. Both undernutrition and malnutrition result in disease. In addition, people who get too little food or nutrients are not as productive as others.

Not all underfed people live in developing countries. Many people in the United States are malnourished. They eat foods that do not supply enough nutrients. This may be due to a lack of thought, a lack of education, or sometimes, poverty. Around the globe, poverty is the cause of most hunger.

Many people do not have enough money to buy sufficient food and lack the resources to produce it for themselves. Hunger can cause people to be unable to work productively. It can cause children to be unable to concentrate and to do poorly in school. People

Many food agencies provide food for countries that cannot produce the right kinds of food.

Rice fields are a common site in China and throughout Asia.

who are hungry most of the time do not have much of a chance to get out of the **poverty cycle**.

Is There Enough Food?

As you've read, the problem right now is not so much in having enough food in total, but of each person getting enough. Two of the largest countries in the world, China and India, now produce enough grain to feed their own people, but can they continue to do so? Both India and China are currently overpumping their water supplies in order to keep up with their irrigation needs. Asia has over half of the world's population and currently produces 90% of its rice. However, technological advances for increasing rice yields have almost stopped since 1985. In Asia, the rate of increase in rice production is beginning to slow down.

On a worldwide basis, population is increasing at the rate of less than 2% per year. At the same time, food production has been increasing at the rate of 3% per year. But statistics such as these can be misleading. Much of the increased food production has come from increasing the amount of land devoted to farming. Now there is not much more land available that can be turned into farmland. It is estimated that in this decade, the land available per person to produce our basic food needs will decrease by 1% per year.

Industrialized farming is one way of increasing yields on available farmland.

Fertilizers promote the growth of larger, more robust plants. Chemicals to control insects and other pests give crops an additional advantage. Tractors and other farm equipment can also help a farmer tend more land. But each of these solutions is expensive in terms of money and the damage they do to the environment. Developing countries cannot obtain the same amount of expensive chemicals and machines, and producing the chemicals and machines, and running the machines, requires fossil fuels, which are in limited supply.

Farming Techniques

In the 1950s and 1960s, some developing countries started using more irrigation, fuel, and fertilizer. They also started using improved strains of grains and livestock. The result was so dramatic that it was called the **Green Revolution.** However, in many countries the growth in farming is reaching the limits of the land and water supply. Also, in some countries further technology is not available to raise food output. The Green Revolution was largely due to an enormous increase in the use of chemical fertilizers. Today, many countries have reached the point where use of additional fertilizer cannot really increase crop production further. Also, there are growing concerns about the use of farm-related chemicals and their effects on the environment.

It may be possible to improve crop yields to some degree without using chemicals. A method known as **sustainable farming** seeks to greatly reduce the use of chemical fertilizers, insecticides, and herbicides. Another method of farming called **organic farming** does not use chemicals at all.

More advanced technology can lead to changes in the plants themselves. **Genetic engineering** allows traits from one organism to be implanted in another. For example, wheat plants may be developed that could resist insects and survive periods of drought, thus eliminating the threat of chemicals and diminished water resources.

Governments in developing countries could also help by giving farmers incentives to increase their yields as well. By improving roads and storage areas, food would stay fresh until it was ready to be used.

Food and the Future

Will there be enough food for everybody in the future? The problem is that the world population is continuing to increase and we are quickly approaching the world's maximum food production. As an example of this we can look at **world grain surpluses.** This is the amount of wheat, rice, and corn that is being stored throughout the world. In 1987 these stores equaled 461 million tons. That's enough to feed the world for 102 days. In 1990 grain stores dropped to 290 million tons. That is enough for only 62 days. At the current rate of population growth, the world will need to produce more food in the next 50 years than it did in the last 10,000 years! If populations continue to increase and food production cannot keep pace, some experts predict wide-scale **starvation,** people dying for a lack of food, in the next 30 years.

As you can see, there are many problems associated with our food supply. In order to grow enough food for the world population and to get it to everyone, many issues must be addressed. As with many problems of this nature, there are no simple answers.

DISCOVERY

WHY ARE SO MANY PEOPLE GOING HUNGRY?

Today, more people in the world are hungry than ever before. This is true even though there is enough food produced worldwide to feed everyone. Why, then, do so many people go hungry? To find out, try the following activity.

Materials (per group)

popcorn
juice
bag of numbered tags
paper cups

Procedure

1. Choose a numbered tag from the bag your teacher has prepared.
2. If your tag has the number "1" on it you are a member of group 1. Sit at the table designated by your teacher for Group 1. Group 1 represents the 15% of the world's population living in the richest countries of the world.
3. If your tag has number "2" on it, sit at the table designated by your teacher for Group 2. Group 2 represents the 25% of the world's population living in the middle-income countries of the world.
4. If your tag has the number "3" on it, remove your shoes and sit on the floor in the area designated by your teacher. Group 3 represents the 60% of the world's population living in the poorest countries of the world.
5. Your teacher will serve a "meal" of popcorn and a beverage to each of the three groups. The amount of food each group receives depends on that group's "income."
6. As you eat your "meal," look around the room at the other groups to see what they are eating. Observe the reactions of other group members.

Observations

1. To which group did you belong? What was the general mood among the members of your group during and after the "meal"?
2. What seemed to be the general mood among the members of the other two groups?
3. During the "meal," did you want to take any action to make the situation more fair? If so, what action did you want to take?

Conclusions

1. What do you think could be done both worldwide and locally to reduce the problem of hunger?
2. List some reasons you think might cause some people to go hungry.
3. In this activity, it is assumed that all people who live in wealthy nations have more than enough food to eat. Do you think this is true? Explain your answer.

Look at the map of the world on the next page.

4. In which areas are 40% of the people malnourished? Why do you think so many people are malnourished in these areas?
5. In which areas are less than 25% of the people malnourished? Why do you think the malnutrition rate is lower in these areas?

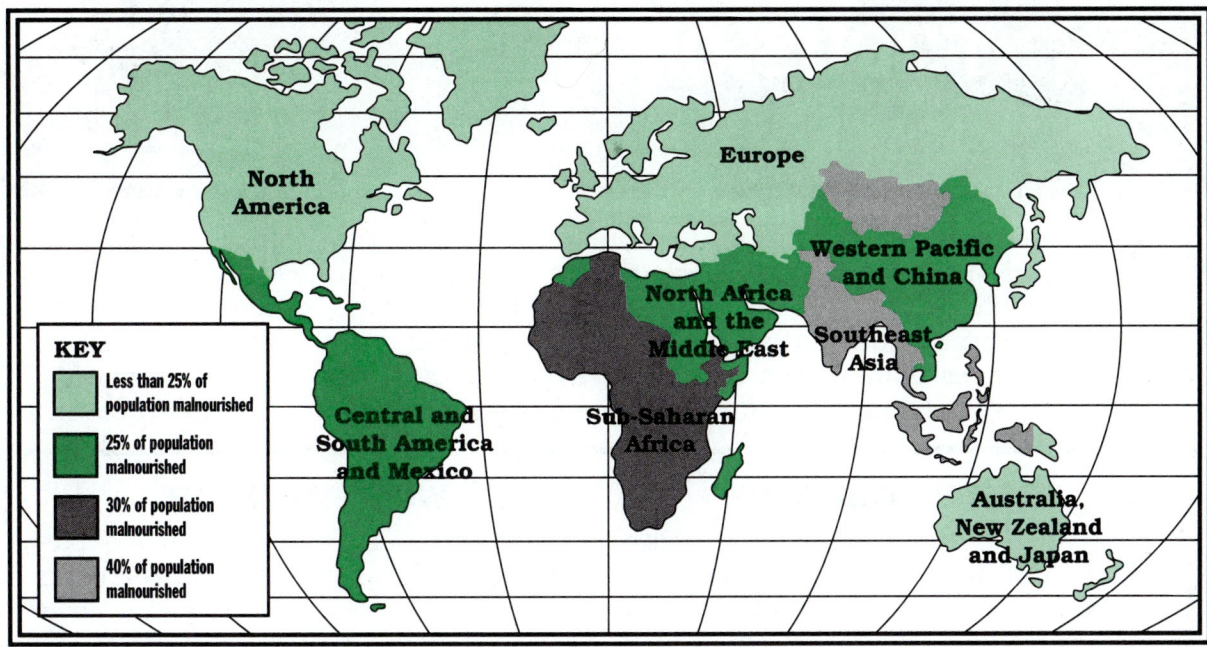

Which areas of the world have the largest percentage of malnourished people?

BRAINSTORMING

Collaborative Learning What are some of the problems or issues that relate to where and how food is grown? What are some issues concerning how food is distributed among and within the different countries of the world? Spend at least five minutes with a group of your classmates making a list of food resource issues. Express each issue as a question. For example, what obligation, if any, do rich countries have to provide poor countries with food, or, how should pesticide use on food plants be regulated?

Remember, when you brainstorm, do not criticize one another's ideas. After your group has completed the list, have one group member report to the class. How many issues did the class come up with? For tips on Brainstorming, refer to the skill on brainstorming on page 231 of this book.

SCRAPBOOK

Gathering Information Clip out newspaper articles that relate to hunger or food-use issues in your community as well as throughout the world. Look in newspapers with local, regional, and national circulations. Continue adding to your scrapbook as long as your class works on this module. For tips on making a scrapbook, refer to page 232 of this book.

JOURNAL WRITING

Critical Thinking When you hear the term *world hunger*, what comes to mind? In your journal, express your thoughts about world hunger and your ideas on how the problem might be solved. As you work on this module, continue to express in your journal your evolving reactions to this world problem. For tips on keeping a journal, refer to page 233 of this book.

FOCUS ON...
1 Worldwide Famine

Starvation. By definition, it is the state of suffering from extreme and prolonged hunger. On the continent of Africa alone, over 21 million people are on the verge of starvation. Worldwide, over half a billion people are in a continual state of hunger. Find out what complicated set of circumstances contributes to hunger in Africa and the rest of the world. Then, try to determine what direction we should take to solve this problem.

BEFORE YOU READ

Read the statements below. Can you think of events that have happened recently that might support these statements? Divide into groups of three or four students and discuss your answers.

CAUSE	EXAMPLES
War can cause hunger problems.	
Major political changes in a country's government affect hunger.	
Natural disasters can cause hunger problems.	

WHILE YOU READ

Copy the chart below onto a separate sheet of paper. As you read the articles, fill in the solutions that have been suggested to fight world hunger. Pair up with a partner and discuss the merits of these solutions. Pick a solution that might help solve a local hunger problem.

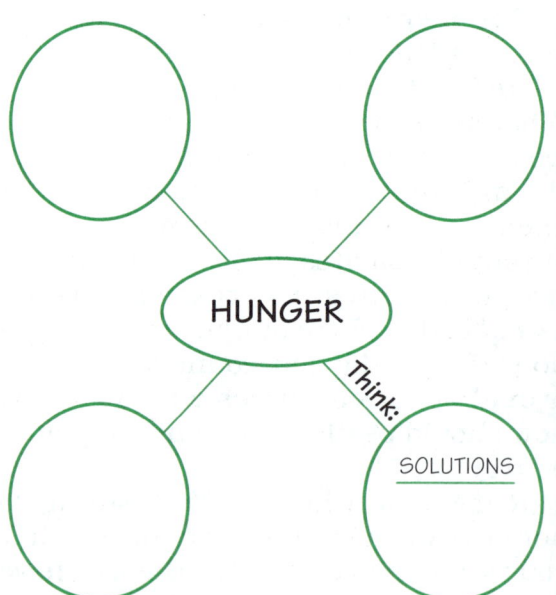

Famine in Africa, The Other Desert Crisis

FROM: *THE ARIZONA REPUBLIC*
MARCH 7, 1991

The Horn of Africa is no cornucopia [horn of plenty]. Far from it. The United Nations Food Program estimates that up to 21 million people there are on the point of starvation.

What makes the tragedy worse is that natural causes are not to blame. East Africa is experiencing a drought [period of dryness], which compounds the situation to be sure, but the real culprit [thing guilty of fault] is politics.

The people of Ethiopia, the Sudan and Somalia, all ravaged [destroyed] by civil war, are suffering because they have been all but abandoned to hunger and starvation by leaders obsessed with the politics of death and destruction. Money is spent to wage civil wars, not fight the famine.

Contemplate [consider] for a moment the grim picture of war and famine painted by *Bread for the World,* a Washington-based anti-hunger lobby. In the Sudan an estimated 11 million people are imperiled [in danger]. It is calculated that the civil war is costing the Sudanese government about $1 million a day. At the same time, the government rebuffs [rejects] U.N. relief operations designed to get food to its starving people.

In Ethiopia, where more than 5 million people are starving, the government is spending approximately 70 percent of its budget for military and security purposes.

In Somalia 7 million people are the victims of civil war and famine. It is estimated that more than 50,000 people have been killed by the Somalian government in the past two years.

Both governments and rebel forces on the African Horn use food as a weap-

Five million people are starving in Ethiopia, while 70% of the country's budget is being used for the military.

on, refusing to let humanitarian aid pass into opposition-held areas. The wars also disrupt government projects —such as they are, since armaments [military equipment] take precedence over development—and displace people from ancestral [inherited from family] lands.

Bread for the World draws a striking parallel between the "other" desert crisis and the Persian Gulf war triggered by Saddam Hussein's invasion of Kuwait. Like the people of Iraq, those who live in east Africa have been ruled by treacherous [dishonest] leaders who trample on human rights.

Bread for the World makes a strong case for the United States to play a constructive role in promoting peace in the region. This could be done without the use of U.S. military might.

The anti-Iraq gulf coalition showed that the U.S. is able to exert impressive diplomatic power on nations of all ideologies [a body of ideas] in forging [forming] a consensus [general agreement] against "naked aggression." Pressure could be brought to bear, if we chose, to stop arms sales to Horn countries and to encourage warring factions [groups] to seek peaceful solutions. Meanwhile, with millions of lives at stake, relief supplies should be allowed to get through to the starving.

World Hunger Is Persistent But Not Inevitable, Says New Report

By George D. Moffett III
FROM: THE CHRISTIAN SCIENCE MONITOR
THURSDAY, OCTOBER 17, 1991

Few regions of the world are as inhospitable as the African Sahel [region south of the Sahara], where the relentlessly [not stopping in intensity] expanding desert has consumed vast agricultural lands and created persistent [existing for a long time] famine.

Twenty years ago groups of young villagers in Burkina Faso decided to fight back. With almost no money and only the simplest of technologies, they began building primitive dikes to trap scarce rain water long enough to moisten small plots of land. The result was a significant increase in crop yields. Long-term effects have been even more promising: reclamation [recovery] of hundreds of acres of farmland once lost to the encroaching sand.

The work of the so-called "Naam movement" in Burkina Faso is one reason that the percentage of the world's population that is hungry is slowly declining, according to a report issued yesterday by the *Bread for the World* Institute on Hunger and Development.

But even as the percentage declined, the report concludes, the absolute number of hungry people continued to grow in 1990–91.

Food shortages are most acute in the Asia/Pacific area, where a majority of the world's hungry now live. But hunger is worsening even in bastions of prosperity like the United States, where the number of people living in poverty rose from 31.5 million in 1989 to 33.6 million in 1990.

The report says several special circumstances exacerbated [worsened] food shortages over the past year, including the Persian Gulf war and sweeping political change in the former Soviet bloc. Caught between economic systems following the collapse of communism, 80 million people in the Soviet Union now are vulnerable [susceptible] to hunger, according to the report—entitled "Hunger 1992."

Also, civil wars in Africa have destroyed agricultural land and disrupted transport and marketing.

In all, half a billion adults and children are in a continual state of hunger, while another half billion are too poor to obtain an adequate diet for a productive work life. The two categories represent 20 percent of the world's population.

A study released Tuesday [October 15, 1991] by the United Nations Food and Agriculture Organization adds that food shortages will grow worse as world population grows from 5 billion to 8.5 billion over the next 30 years, and as deforestation diminishes arable land by 42 million acres each year.

Despite these grim statistics, massive hunger is not inevitable [certain], the *Bread for the World* report concludes. "The principal barrier to ending world hunger is neither lack of resources nor insufficient knowledge," says *Bread for the World* president David Beckman. "It is the failure to put ideas that work into practice on a broad scale."

One idea that works is the participation of hungry people—like the villagers in Burkina Faso—in planning and implementing local projects to increase food production. "Because hunger results from a complex set of factors relat-

Somalian refugees driven into eastern Ethiopia by lack of food. Some victims of poverty have been fighting back.

ed to poverty and powerlessness, successful efforts to reduce hunger must involve the intended beneficiaries [people who benefit from something] in making decisions," says the 200-page report.

Another idea that works is careful targeting of aid, like food recently donated by the U.S. that tided over [helped through a difficult time] a group of Indian workers who moved from Bombay to rural Maharashtra to begin self-sufficient farming.

Government policies that contribute to food self-sufficiency also help combat shortages. In Indonesia, for example, smart economic management coupled with strong investment in the rural sector has spurred farm output and contributed to a dramatic reduction (from 60 percent in 1970 to 17 percent in 1987) in the number of people living below the poverty line.

The report says that a major cause of hunger is militarization [equipping with military forces], which robs developing nations of the resources needed for food and social services. Developing nations spent nearly $400 billion on arms in 1988.

One success story is Costa Rica, which abolished its army in 1949 and spent the savings on social programs. The result has been a higher level of social services and fewer people below the poverty line than virtually [almost entirely] any other country in Latin America.

"They took the money that would have gone into military spending, and it's paid off in lower levels of hunger and poverty," says Mr. Beckman.

These people are receiving fresh water, something rare in the Iraqi countryside.

In A Changing World, Little Has Changed For The Hungry

By Reena Shah
FROM: THE ST. PETERSBURG TIMES
OCTOBER 16, 1991

The world grows enough food for everybody, but the number of hungry people remains unchanged: One out of every five people can't afford the basic amount of food needed to maintain an active work life.

Natural disasters and man-made problems aggravated [made worse] hunger throughout the world, according to a report released by an advocacy group to mark World Hunger Day. Floods ruined harvests in China and Bangladesh. Drought combined with civil war uprooted more than 22 million people in Ethiopia, Sudan, and Somalia. The Persian Gulf war subjected Iraq—including its Kurds and Shite Moslems—and foreign workers from Asia and Africa to severe food shortages.

The breakdown of the political and economic systems of the Soviet Union disrupted food supplies, hurting its people even more because the majority live below the poverty level. And change in the Soviet Union affected countries that relied on the Soviets: It caused cuts in aid to countries like Cuba and Vietnam, causing food shortages there.

Even the United States was affected. The number of people collecting food stamps because they lost jobs and incomes to the recession [temporary decline of business activity] increased to 23 million. That still didn't cover all the country's poor; 32 million people live below the U.S. poverty level.

> **"It's not that we don't have the solutions. All we have to do is put it into action."**

All this isn't cause to lose hope, though, said Don Reeves, director of *Bread for the World,* the Washington-based non-profit lobbying group that issued the report.

"More people were hungry, but more people also enjoyed a better level of nutrition," Reeves said. "It's not that we don't have the solutions. We have the know-how. All we have to do is put it into action."

Governments need the political will for change, Reeves said. They need to enable poor people to get the technology and credit for self-help. And, he said, they ought to cut military spending: Currently, the amount the world spends on weapons exceeds the income of half its 5 billion people.

Food aid from the United States helps people in poor countries, but less grand measures could help them more, Reeves said. For example, if the United States cuts its budget deficit [negative balance of funds], it could help small farmers in other countries far more than foreign aid. "We're sucking up so much of the credit that's available that we're pushing up interest rates for everybody else," Reeves said. If the United States were to help cut world interest rates by 1 percent it could save indebted countries about $6 billion in interest payments. That's four times the amount of help the United States gives in food aid.

1 GETTING INVOLVED

Gathering Information

Write a Letter to Request Information One point of view presented in these articles is that funds should be rechanneled from other areas and used to supply food to those people in need. If you agree with this approach, write a letter to the *Bread for the World* organization to find out what can be done at the local level to make sure this happens. The address for this organization can be found in the Community Resource Directory, starting on page 244. Share this information with the rest of your class.

Critical Thinking

Investigate Locally Contact an organization such as the Salvation Army or a local soup kitchen. Find out what they are doing to feed those in need in your area. Ask them for data about how many individuals they help. Find out where these organizations get the money to provide food for the hungry. Do they have enough money and support to feed all the people who suffer from hunger in your community? Share this information with the rest of your class. Were your classmates aware that people in your community are suffering from hunger?

Decision Making

Write a Speech Imagine that you are a member of the United Nations Food Program. You have decided to present a speech to the United Nations that addresses how this organization should approach the problem of world hunger. In your speech, make sure you define the problem, its causes, its effects, and the most effective solution or solutions. Present your speech to your class. Have your classmates evaluate your presentation and the effectiveness of your approach.

2 FOCUS ON...
Hunger in The United States

Hunger is not just a problem of Third World countries such as Ethiopia and India. People in the United States are going hungry, too. A recent study by the Food Research and Action Center focused attention on the 5.5 million young children in this country who are suffering from hunger. Read the following two articles about the side effects of hunger. Then find out what is being done to solve this problem, which affects a large proportion of the nation's poor.

BEFORE YOU READ

Think about what you already know about hunger. How does it affect a person's daily life? Sketch a chart similar to the one shown below. In each circle, describe how hunger over long periods might affect these specific areas.

WHILE YOU READ

While reading the articles, jot down the specifics of the Mickey Leland Bill. Imagine you are a representative determined to get the bill passed. Using the outline below, prepare a statement supporting your cause.

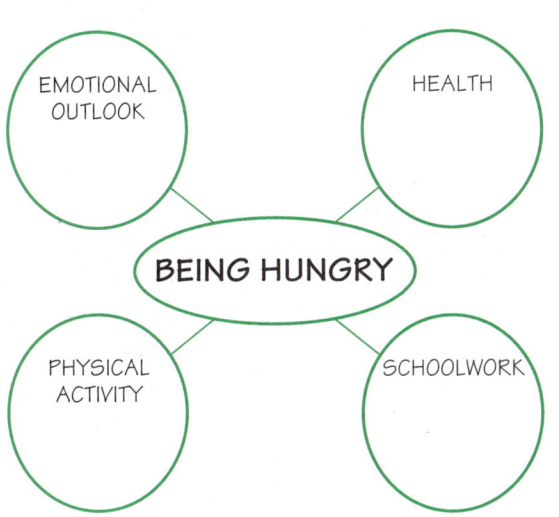

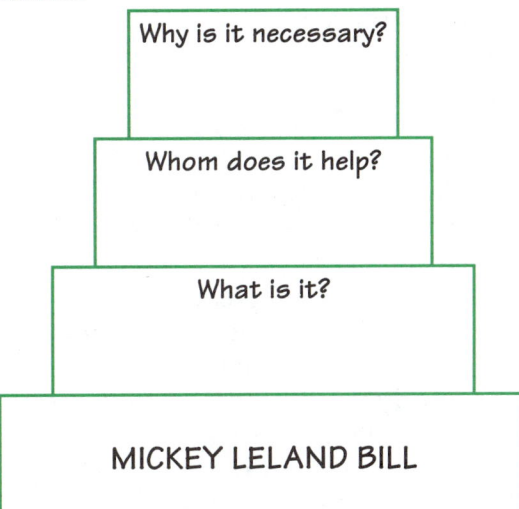

Hunger Said to Afflict 1 in 8 American Children

Millions More on Verge, Group Estimates

By Spencer Rich
FROM: *THE WASHINGTON POST*
MARCH 27, 1991

One of eight children in the United States under age 12 suffers from hunger, and millions more are in danger of going hungry, according to a new study that sponsors said is one of the most comprehensive [thorough and wide-ranging] investigations of the problem.

"The ultimate tragedy of hunger is that it is wholly preventable. We have more than enough food in this country to ensure that every child is properly fed," said Robert J. Fersh, executive director of the Food Research and Action Center [FRAC], the advocacy group that conducted the survey.

The study found that 5.5 million children under age 12 are hungry, and that millions more are "at risk," [said Cheryl Wehler, who headed the study]. . . .

The group [FRAC] interviewed 2,335 families at seven sites, both urban and rural, over the past two years. Families were asked whether lack of money or government aid had forced them within the preceding year to cut food portions, send the children to bed hungry, limit the number of foods served, and give children meals they clearly knew were not adequate. If they answered yes on at least five of eight such key questions, the children were classified as hungry. A large proportion were hungry every month for at least a week, Wehler said.

The study was announced as Fersh's group and about 90 other welfare, religious, and anti-hunger groups began a drive to pass the Mickey Leland Childhood Hunger Relief Act. The bill, named after the late congressman from Texas, would change the food stamp program to make sure that families get more food aid. They are also seeking improvements in school food programs, especially school breakfasts, and more funds for the special food program for low-income pregnant women, infants, and children (WIC) so that it covers all eligible families. . . .

Fersh said that the WIC program serves only about half of those eligible because of funding limits, and food stamps reach only 50 percent to 60 percent because many people don't know they are eligible, among other obstacles [something that prevents action].

The Bush administration has proposed increases for some programs, and Fersh says they are constructive [helpful] but not sufficient [enough]. For example, he said, his group favors increasing WIC funds enough to provide full funding within five years for all people

HUNGER: HOW WIDESPREAD A DANGER?

PERCENTAGE OF CHILDREN UNDER 12 WHO ARE HUNGRY OR AT RISK

STATE	HUNGRY	AT RISK	STATE	HUNGRY	AT RISK	STATE	HUNGRY	AT RISK
Alabama	17.0%	18.0%	Kentucky	14.3%	15.5%	North Dakota	12.8%	15.1%
Alaska	8.3	9.4	Louisiana	15.9	16.2	Ohio	11.7	12.6
Arizona	12.1	13.7	Maine	11.2	13.1	Oklahoma	13.5	15.2
Arkansas	18.4	20.1	Maryland	7.9	8.5	Oregon	11.3	12.8
California	13.1	14.7	Massachusetts	9.3	10.0	Pennsylvania	12.3	13.3
Colorado	9.9	11.1	Michigan	13.3	14.3	Rhode Island	12.2	13.4
Connecticut	7.8	8.4	Minnesota	9.8	11.4	South Carolina	15.0	16.3
Delaware	10.6	11.2	Mississippi	18.9	19.8	South Dakota	13.4	15.2
D.C.	17.1	16.6	Missouri	12.7	14.2	Tennessee	14.7	16.0
Florida	13.2	14.3	Montana	14.7	16.8	Texas	13.6	15.0
Georgia	13.1	14.0	Nebraska	13.2	15.4	Utah	11.3	14.2
Hawaii	10.8	12.9	Nevada	12.9	15.0	Vermont	10.0	10.5
Idaho	15.6	18.8	New Hampshire	5.8	6.8	Virginia	9.7	10.8
Illinois	13.9	14.1	New Jersey	10.0	10.3	Washington	10.7	12.0
Indiana	12.0	13.3	New Mexico	17.2	18.9	West Virginia	18.3	20.3
Iowa	14.7	17.0	New York	14.6	15.1	Wisconsin	11.4	13.3
Kansas	10.5	12.3	North Carolina	12.4	13.9	Wyoming	10.2	12.4
						UNITED STATES	12.8%	14.0%

NOTE: Percentage "at risk" represents children living in households that experienced aspects of hunger.

SOURCE: Food Research and Action Center

eligible, but at the rate of increase proposed by Bush for 1992, it would take 15 to 20 years.

Fersh's recommendations would add about $15 billion to the $25 billion to $30 billion the government now spends annually to combat [fight] hunger.

Victor Sidel, a physician and professor of social medicine at Albert Einstein College of Medicine, said the survey was not designed to measure malnutrition, which has clear symptoms.

Sidel said the "subtle effect of not getting enough food happens long before we ever see poor growth."

He said [malnourished] children often don't have the energy to learn, and as a result, don't do as well on tests and can be disruptive [cause trouble] in a classroom. They also are less resistant to illness and more likely to miss school, he said.

The survey found that of the families interviewed that were experiencing hunger, the majority were poor, even though many had incomes from jobs. The poorest households surveyed spent 60 percent of their incomes on housing, which cut into food budgets. Among those below the poverty line, average outlay per meal per person was only 68 cents.

Representative Thomas J. Downey (D-New York) chairman of the House Ways and Means subcommittee on human resources, said, "These sobering statistics confirm that we must do a better job nurturing America's children." He said it appears that House leaders will endorse a "bold initiative" [ambitious push] in favor of the Leland act, and a major child welfare bill.

Survey: 160,000 Georgia Children Go Hungry

Organizations urge more federal spending on government food programs—especially for school breakfast programs.

By Rebecca Perl
FROM: *ATLANTA JOURNAL*
MARCH 27, 1991

More than 5 ½ million American children—including one in eight Georgia kids—are hungry and regularly go without food, according to the most comprehensive [complete] survey yet of hunger among poor families in the United States.

There are at least 160,000 hungry children in Georgia alone, despite government and charitable programs over the past three decades aimed at alleviating [lessening] hunger among low-income families, according to reports released Tuesday by the Atlanta Community Food Bank and the Washington-based Food Research and Action Center.

The organizations called for more federal spending on government food programs—especially for school breakfast programs.

Both state and U.S. estimates of hunger are based on a three-year, $1 million national study that included a door-to-door survey of 2,335 low-income families with children under 12 in seven states, including Alabama and Florida.

A SAMPLING OF SOUTHERN RESPONSES

Researchers surveyed 366 low-income families with children under 12 in Sumter County, Alabama, and 273 in Polk County, Florida, and found:

▶ 70 percent of poor Sumter County families say they regularly run out of money to buy food.
▶ 12 percent said children in their Sumter County families go to bed hungry, on average, more than six days each month.
▶ Nearly 30 percent of Sumter County families said they cut the size of their children's meal or their children skipped meals because they didn't have enough money for food.
▶ Nearly half of those eligible for food stamps in Polk County did not get them, mostly because they didn't know they were eligible.
▶ The majority of Polk County families who did get food stamps said they did not last through the month.

Source: Food Research and Action Center
Adapted from: *Atlanta Journal*

The researchers asked questions of families whose annual income is less than 185 percent of the federal poverty level—the level used to determine eligibility for free school lunches and food stamps. (Poverty level for a family of four was $12,700 in 1990. The survey included families making less than $23,495 a year.)

A poor diet leads to stunted [much smaller than normal] growth, learning and behavior problems, and an increased susceptibility [openness] to lead poisoning and anemia [shortage of red blood cells], child health experts say.

Hungry children are twice as likely to have ear infections, colds, headaches, and miss school, the study found.

Overall, they suffer from two to three times as many health problems—including exhaustion, dizziness, irritability and inability to concentrate—as those who get enough to eat.

"We have plenty of food, and kids learn better when they eat," said Bill Bolling, director of the Atlanta food bank, which collects and distributes food to organizations that feed the hungry in metro Atlanta and North Georgia. "They concentrate better if they are fed. We know how to do it, we have to do it, it's just common sense."

Federal food assistance programs—school breakfast, lunch, and summer feeding programs and food stamp and nutrition programs—are falling short all over the United States, including Georgia, state advocates say. . . .

Even though the federal government pays for school breakfast programs for low-income children, some local school districts find it inconvenient to juggle bus and work schedules so they don't take advantage of them.

"And some have the basic belief that parents ought to feed their kids in the morning themselves," Mr. Bolling said.

Only 4 percent of Cobb County schools offer breakfast programs; 14 percent of schools in Gwinnett and 23 percent in DeKalb [offer them], advocates say.

Federal funds for other programs are simply inadequate [not enough], the researchers said.

Girl Scouts collecting food for a local Food Research and Action Center.

For instance, a supplemental [added] food program that offers milk, eggs, juice, and bread to women, infants, and children, called WIC, has only enough money to serve 46 percent of those eligible in Georgia.

Because there was not enough federal money last year, some Georgia women with newborns had to wait nine months to get into the program, say those who administer it.

Almost 60 percent of Georgians eligible for food stamps don't get them, largely because they don't have the time or transportation to pick them up at food stamp offices, local officials said. Illiteracy [inability to read] is also a barrier to the program applications and rules.

2 GETTING INVOLVED

Gathering Information

Find Out More Go to the school or local library and research hunger. Find out what some of the effects of hunger are and how they affect the health and well-being of children. Find out what programs are available to help people who do not get enough food. Summarize your findings in a brief report.

Decision Making

Defend Your Decision Now that you have read the two articles, make your own decision about whether breakfast and lunch programs are the answer to preventing hunger in children in the United States. Think about the problem and its causes and effects. Will a breakfast and/or lunch program in every school meet the needs of all school-aged children? What about children who are not in school yet? What about families who do not take advantage of the program? Defend your answer with facts from this and other articles that you can locate. If you do not think that these programs will solve this problem, offer a better solution.

Critical Thinking

Investigate Locally As a class, find out about the different agencies in your area that may deal with the issue of hunger. One group of students may want to contact government agencies, such as the community welfare department. Another group of students may want to contact religious interfaith groups. A third group of students may want to contact non-profit agencies, such as the United Way. Each group should find out the following information:

1. Agency name and address
2. Agency contact person's name
3. Services that agency provides
4. The amount of money that agency spends on salaries, and how much goes into program.

After the class has gathered the information, decide on an action the class may want to take to help the hungry in your area.

3 FOCUS ON...
Two Approaches to Farming

As concern about the use of farm chemicals and their effects on the environment grows, people are looking for ways to reduce their use and still meet our growing need for a plentiful food supply. As a result, two opposing approaches to agriculture have emerged and are clamoring for support. In organic agriculture absolutely no chemical or non-natural pesticides, herbicides, or fertilizers are used. In sustainable agriculture, on the other hand, farmers use such chemicals, to ensure a plentiful and high-quality food supply. Which method would you support? Make your decision after reading the following two news stories.

BEFORE YOU READ

Look at the word groups below and for each one decide which word does not belong. Why? What is the relationship among the other words?

Word Groups
1. chemical, herbicide, pesticide, natural
2. nutritious, healthy, soil fertility, contaminated
3. recycling, ecology, disease, environment
4. fertilizers, insecticides, synthetic, organic

WHILE YOU READ

Copy the chart shown below onto a sheet of paper. As you read the articles, list the factors that support sustained agriculture and those that support organic agriculture. Now imagine you are a new farmer trying to decide which method of agriculture to use. Which factor is most important to you? Be prepared to defend your answer.

Sustainable Agriculture | Organic Agriculture

Statement that is of greatest concern to you: _____

Sustainable Agriculture More Pragmatic Than Organic Farming

By James A. Carlson
Associated Press
FROM: ST. PAUL PIONEER PRESS
JANUARY 17, 1991

Rural sociologist Peter Nowak sees a big difference between sustainable agriculture and the organic farming popularized [made popular] by the hippie [young people who were against the laws and rules of society] culture of the 1960s. . . .

Organic farming is based on a strict ideological approach that no commercial pesticides or herbicides should be used, while sustainable agriculture focuses on efficiency and improvement through flexibility.

Often that means reducing reliance [dependence] on chemicals, but it is not a rigid rule. Sustainable farmers are flexible to adapt to changes in the weather, the market, or technology.

"The farmer does not get locked into a pattern where there is only one way for him to go," Nowak said. "A sustainable farmer goes with the flow."

If hit with an outbreak of a plant disease, a sustainable farmer can use chemicals necessary to cure it. But the sustainable farmer also uses crop rotations, tillage [plowing] methods and other practices that reduce reliance on commercial chemicals.

In the decades before the sustainable agriculture movement in the mid-1980s, the farming industry became reliant on technology and scientific advances.

Farmers in those days spent less time concerned about their land and pollution and more time finding quicker, easier ways to farm using equipment and chemicals.

Sustainable agriculture advocates say [that today] their practices emphasize using practices that better utilize natural resources through crop rotation, tillage and manure fertilizing.

"What we're trying to do is expand the range of alternatives, rather than what we've done for the past years so successfully, (telling farmers) this is the one way to raise corn," Nowak said.

Terry Gips, president of the International Alliance for Sustainable Agriculture, said contributing factors in the movement included:

- The farm crisis of the mid-1980s, when debt-ridden farmers faced with possible loss of their land needed ways of cutting expenses.
- Groundwater contamination problems involving such things as the corn herbicide atrazine, the potato pesticide aldicarb, and nitrates from fertilizer.

- Consumer concern about pesticide residue [remainder] in food, as exemplified [shown] by the 1989 scare involving Alar, a chemical used on apples to improve color and shelf life.

- Worsening soil erosion problems that many blame on continuous planting in one crop, such as corn, and the failure to practice crop rotation, strip-cropping, and other methods.

Organic Farming Is The Solution To Residue Pesticides, Expert Says

By Reggie McLeod
FROM: *ST. PAUL PIONEER PRESS*
FEBRUARY 25, 1991

Organic farmers should focus on selling a system of agriculture that benefits the environment and society rather than simply claiming that organically grown food is more nutritious and contains less chemical residue, an international consultant on ecological agriculture told a group of 100 farmers Saturday.

"We all know that organic farming is the long-term solution to pesticide residue problems," Joe Smillie said. "Our food is produced by a farming system that produces land health, community health, and planetary health."

Smillie, the keynote speaker at the three-day Wisconsin Organic Farming Conference, is a founding member of the Organic Foods Production Association of North America and author of several books on organic agriculture.

The Canadian works with organic growers, marketers, certifiers, and processors around the world....

Smillie, in his keynote speech, said that all food, whether organically grown or grown with chemicals, contains some pesticide and herbicide contamination [impurity] because these chemicals are now present in varying amounts nearly everywhere on the planet. Organically grown crops usually contain far less residue, but it is difficult to get that message across to consumers. There is an ongoing debate as to whether organically grown crops are more nutritious.

The clearest benefit of organic farming is that it does not put chemicals into the soil, air, and water, he said. However, he said, that is difficult to market. Is a consumer likely to think, "I will eat this and improve the world"? Smillie asked.

Another current problem with organically grown crops is that there is no national labeling standard, although some [states], such as Minnesota, have set minimum standards. National standards may be in the final version of the farm bill now working its way through Congress, he said. . . .

Sustainable agriculture aims at reducing inputs of energy, petrochemicals, and other chemicals in farming. Organic farming, on the other hand, does not use any of an array of chemical fertilizers, insecticides, herbicides, or other synthetic chemicals.

. . . Many farmers get interested in sustainable agriculture to cut costs, then become interested in the ecological impact of farming, he said.

3 GETTING INVOLVED

Cooperative Learning

Debate the Issue Work with four other students to debate the issue of organic vs. sustainable agriculture. One person should be the moderator, two students should take the side of organic farming, and two students should take the side of sustainable agriculture. Using the information presented in the articles, as well as your own opinion, prepare a one-minute statement, concerning organic or sustainable farming. Develop at least three questions to ask the opposing group. After each opening statement has been given, continue to debate the issue. After ten minutes, the moderator should summarize what was said and decide which side presented the most convincing argument.

Gathering Information

Investigate Locally Visit your local hardware store, or discount department store. Locate those products used for lawn and garden pest control. Make a list of these products and identify their uses and their hazards. Pay particular attention to the warning labels on these products. Then, contact the Environmental Protection Agency (EPA) and find out if there are safe alternatives to these chemicals. In addition, find out if there are any pest control chemicals that have been recently banned in your state. (The phone number and address for the EPA can be found in the Community Resource Directory starting on page 244.) For some tips on using the telephone effectively, refer to page 235.

Problem Solving

Write a Public Service Announcement A public service announcement (PSA) is information that is broadcast free of charge because the information is considered to be for the general good of the public. Use what you have learned about organic farming to prepare a public service announcement. Include the benefits of organically-grown food over food that is grown using chemicals. Ask your principal if you can broadcast your announcement over your school's public address system.

4 FOCUS ON...
Threat to California Rice Farmers

For 80 years, rice has been a major crop in the Sacramento Valley. It is one of California's oldest crops. But today the future of this crop is being threatened because of growing demands for the water that is needed to flood the rice fields. Find out how population growth and drought have combined to threaten the livelihood of some of California's oldest farm families. Then think about what you would offer as possible solutions to this growing problem.

BEFORE YOU READ

Copy the chart shown below onto a separate sheet of paper. Work with a partner and brainstorm everything you know about rice. Fill in your answers on your chart.

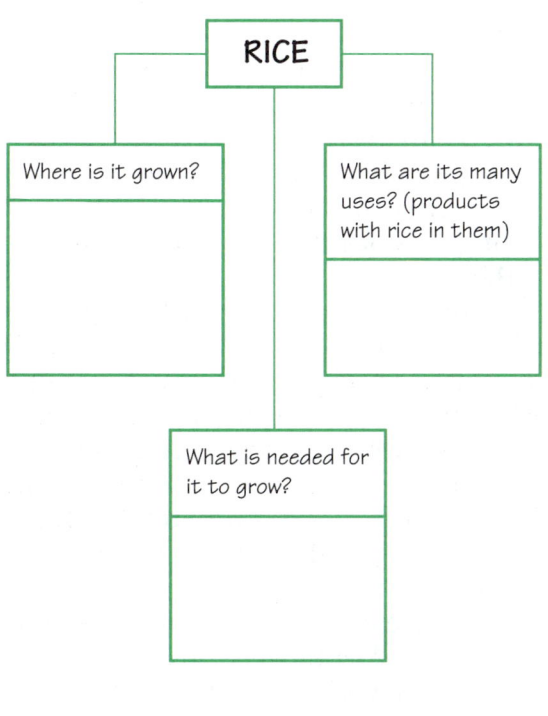

WHILE YOU READ

Copy the chart shown below onto a separate sheet of paper. While you are reading the article, pay close attention to the concerns of those groups who do not want farmers to continue growing rice in California and to the views of the rice farmer. Record the views of each group in the appropriate space in your chart. After completing the chart, consider all of the arguments. Which side do you feel has the strongest argument? Explain why?

Growing Rice in the Desert		
	PRO	CON
Water Issue		
Population		
Environmental Issues		
History/Economics		

State's Growth Threatens Way of Life in Rice Towns

Paddies lose more water to evaporation than L.A. uses in a year

**By Kevin Roderick,
Times Staff Writer**
FROM: *THE LOS ANGELES TIMES*
APRIL 7, 1991

A rattlesnake skin hangs behind the men's toilet at Richvale Cafe. Swedish travel posters spiff up the lunchroom. Blueberry pie baked by somebody's mom awaits for dessert.

Richvale's lone cafe is run by rice farmers, for the noontime pleasure of rice farmers, in the heart of California rice land. Rice has been sown here in Butte County since 1912, long enough for droughts to be accepted as part of the deal that farmers cut with nature.

This lunch hour, a more bewildering force than drought, newer and more threatening, has Don Murphy poking the air with his soup spoon.

A burly man in low-flying jeans, Murphy wants it known that survival of a classic California culture, the rice towns of the Sacramento Valley, is at risk from the state's population explosion.

Even if the drought ends tomorrow, he says, California is so engorged with people that water will always be in short supply. The cities, slaves to their growth, are making noises about commandeering [taking forcible possession of] the water used on rice farms for three generations. But Murphy pleads it would be a mistake.

"You can't have the country—the means of production—dry up and still keep the cities thriving," said the sharp-spoken Murphy, who put his children through college on [the income from] rice land in his family since the 1920s.

Chairs shift in the cafe; eyes sweep the room. Growth in metropolitan California is an uncomfortable topic in tiny Richvale, now that the forces of population and drought have merged to frame a social question that is a lunch-spoiler in cafes up the Sacramento Valley:

Does rice, which loses more water to evaporation than Los Angeles uses in a year, have a place anymore in a state of 30 million people?

The case against rice, in its simplest form, is this: It drains too much water from the same reservoirs that supply most California cities, for a crop less valuable to the state's economy than turkeys or broccoli.

Cotton, pasture, and alfalfa take more water each year than rice. But rice is the only crop raised on fields flooded with water several inches deep through the hot months of spring and summer.

"It's a monsoon crop in a desert state," says Marc Reisner, a San Francisco author who has stated the case against rice most clearly and most often.

Rice is also a sore point for environmental groups, which covet [want] the water as a way to meet nature's needs—to flush out San Francisco Bay, nurture the dwindling salmon runs, and keep some historic wetlands alive. Pollution from pesticides and field burning, and federal payments that encourage rice growing, rankle even more.

"We should shift water from overproduction of rice to meet higher priority needs," said Corey Brown, general counsel for the Planning and Conservation League, an alliance of 120 environmental groups. "It's just an outrageous situation."

But in towns along the Sacramento River, the state's mightiest natural waterway, rice is king. No region of California has its history so closely entwined [twisted together] with a single crop. . . .

In rice towns like Richvale and Colusa, the idea that melting snow runoff siphoned onto fields for decades might now be put to better use in suburbs like Hesperia and Moreno Valley would be preposterous if it wasn't so painful.

"It's taking water out of fragile rural economies to build more cracker-box houses, bring in more immigrants," said Doug McGeoghegan, manager of the 2,300-acre Gunnersfield Ranch, a rice farm and duck-hunting outpost near Maxwell.

"You put three feet of water on an acre of rice in California, you're going to get 9,000 pounds of a nutritious food," said McGeoghegan, whose family homesteaded in Colusa County in 1876. "Put that water on lawns, what will you get?"

In the town of Williams, 10 varieties of packaged rice and six sacks of bulk rice are on display . . . , pitched at motorists lured off Interstate 5. There is lots of unsolicited comment about rice's attributes as food.

"Ninety percent of the world's population eats rice," said Bill Huffman, spokesman for Farmer's Rice Cooperative and grandson of a Butte County rice pioneer. "They don't eat garlic or cherries."

In the rice-eating world, California's crop is of no special importance. Arkansas produces more, so do Thailand, Brazil, and a dozen nations. About a third of California's rice goes into beer and [cereal] products . . . , a third goes to stores and institutional kitchens, and the rest is exported.

California's rice is coveted most by the Japanese, who prefer the sticky, brilliant-white medium grain not favored by American palates. But California rice is banned in Japan, to insulate [prevent] that country's growers from market forces.

Only 2.1 million people lived in California when the first rice crop was harvested here—fewer than in Georgia or Tennessee at the time. Water was abundant and free, or close to it, as it fell from the Sierra Nevada toward the Pacific.

Dry summers and dense clay soil made the Sacramento Valley ideal—

RAISING RICE IN CALIFORNIA

Five counties along the Sacramento River grow 80% of California's rice, one of the state's oldest crops. But as urban population grows and the drought persists, some critics say water devoted to raising rice would be better used in cities.

KEY FACTS ABOUT RICE

■ **History:** Rice has been grown commercially in California since 1912. Two dams, Shasta (completed in 1945) and Oroville (1968), deliver the water that has helped the rice industry flourish.

■ **Type:** Almost 90% of California rice is sticky medium grain Japonica, prefered by Japanese and Koreans but not most Americans. About 5% is long grain Indica, like that sold as instant rice. About 5% is short grain used in rice cakes and other products.

■ **Amount:** Acreage in California peaked at 625,000 in the early 1980s. Last year rice was cultivated on 385,000 acres, more than any crop except cotton, hay and alfalfa, grapes and almonds.

■ **Sales:** Most California rice is sold in the U.S., Puerto Rico, and Guam. In the early 1980s, most was exported; with Korea the main customer. Korea stopped buying rice in 1983; now Turkey and Jordan are the main importers.

Source: Paul Gonzales/The Los Angeles Times

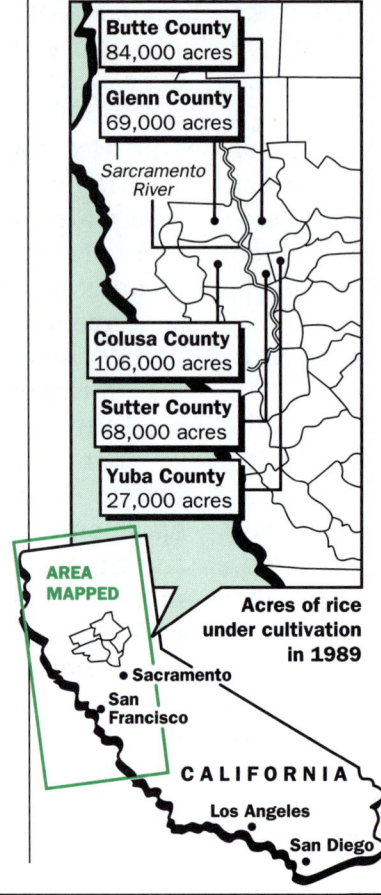

HOW RICE IS GROWN, IN SACRAMENTO VALLEY

1. Fields are leveled, using laser-guided graders, and prepared for planting in winter.

2. Water from the Sacramento, Feather and Yuba rivers is released into fields starting in mid-April. Rice seed is dropped from airplanes, about 150 pounds per acre.

3. Plants are kept under about six inches of water until just before harvest in August. Four to six feet of water must be applied to each acre over the growing season to manage inundation [keeping the land under water].

4. About 3.3 feet of water per acre is lost due to evaporation, seepage, and usage by the plant. In contrast, an average family of five in Los Angeles uses less than one acre-foot a year.

5. Each acre yields 7,000 to 8,000 pounds of rough "paddy rice." Harvest is by combine; rice is one of the most mechanized crops, able to go from seed to final packaging without touching human hands.

6. After harvest, tons of straw are left in the field and usually burned.

7. At mills, the hulls are removed to produce brown rice. Machines scrape off the bran to process the brown rice into white rice. Broken kernels, or brewers' rice, are used in beer brewing.

---

each acre yields more rice than anywhere in the world.

"The Sacramento River Valley is by no stretch a desert," said McGeoghegan, chairman of the Rice Industry Committee, a group campaigning to defend the rice culture. "Rice is grown here for a very specific reason."

Towns sprouted in the valley as rice took hold, giving birth to an infrastructure [underlying framework] of mills and cooperatives. A new co-generation plant in Williams burns waste rice hulls to make electricity. Freighters take on rice bound for Hawaii, Turkey, and Morocco at a river port in West Sacramento.

Colusa, on a lazy river bend 50 miles north of Sacramento, is an outpost of Middle America built with rice money. A stately old courthouse and a square of green lawn anchor the center of town. Graceful shade trees and expensive homes built by old farming families line the streets.

Old rights to the river gave rice farmers a better supply of water than the corporate farms south in the San Joaquin Valley. Then Shasta and Oroville dams were erected to smooth out the river's natural extremes.

Instead of spring floods and summer

trickles, the river delivers a general all-year flow. With the river more reliable, rice growing flourished. But rice also became as beholden to California's water-distribution bureaucracy as any Southern California city.

Water districts have cut deliveries to rice farms in the drought, and rice acreage has fallen from 425,000 in 1988 to about 275,000 this year. Other crops suffered more, but rice began to be talked about as expendable [easy to do without]. This year, state officials came asking rice farmers to sell their water to help cities through the drought.

So far, though, few rice farmers have gone along with the state's request to help stock the "water bank" for cities. Rice uses less water than cotton—and nobody eats cotton, the farmers say.

Lasers that allow precise leveling of fields, and new strains of less-thirsty rice plants, have also made the farms more water efficient. It now takes 25 gallons to grow a serving of rice here, compared to 36 gallons in earlier decades, the industry says.

The water bank has met resistance out of fear it would set a precedent [justify future acts]—that cities could grow at will and later use the power of their superior numbers to tap the rice fields for water. "In a genuine human emergency, absolutely," said McGeoghegan. But not to help Southern California add new suburbs, fill swimming pools and plant gardens.

"Everybody's looking to gore somebody else's ox, it's as simple as that," said McGeoghegan. "Rice looks like an easy mark. It's visible."

In Richvale, up the river in Butte County, a cluster of white-painted elevators and tanks announces Lundberg Family Farms, the area's biggest employer. The family rice business began in 1937 and now includes three brothers and a gaggle of sons, producing mostly organic rice products.

"Water is the biggest concern I have about the long-term viability [survival] of farming," said Bryce Lundberg, 30, the third generation of Lundbergs to grow rice at Richvale. "I would like to stay in the rice industry. My family has built a name. But if you take away the water, you've taken away what makes the land valuable."

His father, Harlan, 57, is a former Peace Corps volunteer who uses a homespun nature to promote the family's product—rice cakes, sweeteners, cereal, and rice for cooking—as health food.

"Rice is one of the premier foods in the world," says Harlan Lundberg. "I don't think there's a food any healthier. It's the major crop in the world, not because it's easy to grow, because it's nutritious."

Harlan says people who think rice competes with the city for water misunderstand nature. The water goes back in the Sacramento River, seeps into the ground, evaporates to become rain again someday, or gets converted into food.

"I don't have much sympathy for those people," he said. "The water—all we do is slow it down a little bit. The mega-problem really is population. There's got to be some way to regulate population."

The tension between rice and growth is not just over water. Environment and health groups have sued the rice industry over the smoke that rises every autumn from burning straw in the rice fields. Studies are trying to learn if the smoke is the cause of an abnormally high lung cancer rate in Sacramento, one of the state's fastest-growing areas.

The cry against rice rose after the 1986 publication of *Cadillac Desert,* an indictment of western water practices by Reisner. He detailed the case again last year in a second book, *Overtapped Oasis.*

Then this year—surprise—Reisner softened his bite, much to the relief of rice growers.

"Irrigating pasture in California is a lot more crazy than rice," said Reisner, a consultant for the Nature Conservancy, who now thinks limited rice growing is good for Northern California.

Little else can grow well in the Sacramento Valley clay, Reisner said, and farmers made a persuasive case to him that the rice fields are good for the dwindling Central Valley winter bird migrations.

"They convinced me the crop is important for waterfowl," Reisner said.

4 GETTING INVOLVED

Gathering Information

Library Research In 1989 rice was grown on about 385,000 acres of land in the Sacramento Valley. Each acre yields between 7,000 and 8,000 pounds of rice. Calculate the annual yield of rice in pounds for the Sacramento Valley. Use a current *World Almanac* to find data about world rice production in pounds. Then figure out what percentage of world rice production comes from the Sacramento Valley. Using this information, do you think the Sacramento Valley produces a significant amount of the world rice crop? Should this information be considered in deciding whether or not rice should be grown in the Sacramento Valley? Why or why not? Summarize your findings in a written report.

Critical Thinking

Prepare a Position Statement Imagine that you are running for political office in the Sacramento Valley in California. You are from a long line of rice farmers, and have decided to run for political office in order to safeguard the future for the rice farmers. Prepare a position statement that includes ways you are going to accomplish this. Make sure you state the problems and your proposed solutions. Explain how you are going to satisfy all the groups opposed to rice farming in the Sacramento Valley, including environmentalists and people in nearby cities who desperately need the water.

Decision Making

Defend Your Decision Now that you have finished reading the article, make your own decision about whether rice should be grown in California. Start by rereading the article and selecting the arguments for and against rice farming. Make sure you separate facts from opinion. Then, prepare a written report supporting one side or the other.

5 FOCUS ON...
Food Irradiation

The word *radioactivity* strikes fear in the hearts of many. Radioactivity is the property of some materials to give off potentially dangerous energy waves. Is it any wonder then, that people are afraid to eat food that has been **irradiated**—that is, intentionally exposed to highly radioactive materials to kill harmful bacteria? Is this fear justified? Do the merits of this process surpass the inherent dangers? Should the nation's first food irradiation plant, ready to begin operation, be allowed to open? Find out why even the scientific community is not in agreement on the answers to these and other questions about food irradiation.

BEFORE YOU READ

Copy the chart shown below onto a sheet of paper. Examine each of the words below. In the column headed "Group 1," list the words most closely associated with the word "radiation." Do the same for "Group 2—Food Safety" and the other group. Use each word only once.

WHILE YOU READ

Copy the chart shown onto a separate sheet of paper. As you read the article, list on the chart the benefits and costs to irradiating foods that are mentioned in the article. Are there more benefits than costs? Which benefits or costs are based on fact and which are based on opinions?

Words

pasteurization	consumer
molds	cobalt
farmers	salmonellosis
radioactive	gamma rays
sterilize	growers
sanitation	market
nuclear	irradiate

Group 1	Group 2	Group 3
Radiation	Food Safety	Food Production

IRRADIATING FOODS	
Advantages	Disadvantages

Irradiated Food Coming, But Not Without Protest

By Larry Rohter, Special to *The New York Times*
FROM: *THE NEW YORK TIMES*
JANUARY 21, 1992

Mulberry, Florida—After years of scientific debate and commercial hesitation, the first food irradiation plant in the nation now stands ready for service and could begin shipping specially treated fruit as early as next week.

Many scientists at universities and federal agencies regard irradiation of food as a safe and efficient way to retard [slow] spoilage and kill organisms that cause illnesses like salmonellosis and diarrhea. Investors in the plant hope that an initial scheduled shipment of irradiated strawberries will soon be followed by other fruits and vegetables and, eventually, poultry and seafood.

"This is going to be a real bonanza for growers and consumers alike," said Sam Whitney, president of the company that operates the plant here. "All the surveys show that people want safer food, and this is a simple, proven process that kills the bacteria that can kill you. It's as important as pasteurization [partial sterilization]."

Arguments About Risk

But opponents argue that "zapping the food supply," as one prominent medical researcher, Dr. Donald B. Louria, has called food irradiation, causes many more problems than it solves. Irradiation not only robs food of some nutritional value and requires the use of dangerous nuclear material, they assert [declare], but may also increase the risk of cancer and birth defects.

"There is enormous controversy over this in the scientific community," said Michael Colby, national director of Food and Water, a consumer advocacy group based in Manhattan that has led the campaign to block food irradiation. "So why are we pushing forward with something unnecessary and easily replaceable? We seem to be recklessly promoting this frivolous [of little importance] technology which is potentially dangerous to human health and threatens the environment."

Three states have acted on the irradiation issue. Maine has banned irradiated produce outright, and New York and New Jersey have imposed moratoriums [legal delays] on its sale.

For many years, limited amounts of spices and edible herbs have undergone the process at some of the 38 commercial irradiation plants around the country, which also sterilize medical equipment and supplies. But the opening of a $7 million plant in this small farming and phosphate mining town 30 miles east of Tampa [Florida] marks the culmination [climax] of an audacious [daring] effort to apply the technology to fresh produce and meat.

Irradiation, a relatively simple process that has been known and studied for 40 years, does not actually make foods radioactive or leave a radioactive residue. Products that are to be irradiated are placed on a conveyer belt, which travels into a chamber that is protected by thick concrete walls, and are then bombarded with gamma rays from a radioactive source like cesium-137, which is a waste byproduct from the manufacture of nuclear power or weapons, or cobalt-60.

Depending on the density and shape of the food product and the amount of cobalt in the 40 wands that rise from a 28-foot-deep pool of water, exposure at the plant here will last 15 to 45 minutes, enough time to kill the insects, molds and bacteria that cause spoilage and disease.

Advocates of irradiation say the process extends the commercial life of fresh foods, enabling growers to reach new markets here and abroad. The strawberries, for example, will have an extended marketing life of up to two weeks, Mr. Whitney said.

Harley Everett, vice president of [the Mulberry company] said, "We're going to start doing tomatoes just as soon as those crops become ripe, and you will see that they will taste just like vine-ripened tomatoes." He also said that irradiation would enable farmers in Florida and other subtropical and tropical areas to grow new crops like wheat without the danger of spoilage and pests, and that yields of some crops like rice would increase for the same reasons.

With an eye on Third World countries, where sanitation is precarious [uncertain or unreliable] and transportation networks are rudimentary [primitive], officials of the World Health Organization and the Food and Agriculture Organization, two United Nations agencies, have endorsed food irradiation. They say the process, which is already in limited use in a score of [20] countries, will also cut down on the consumption of pesticides that damage the environment.

A Government-approved symbol for irradiated food.
Source: Associated Press/The New York Times

Fear of Boycotts

In the United States, the Food and Drug Administration [FDA] approved the irradiation of fruits, vegetables, and grains in the mid-1980's after analyzing 441 studies, and in 1990 it extended that certification to poultry. The American military has also expressed interest in the process, which it sees as a way to provide better tasting and more nutritious food to troops in the field.

But food companies and supermarket chains have been reluctant to introduce irradiated products into stores. That is

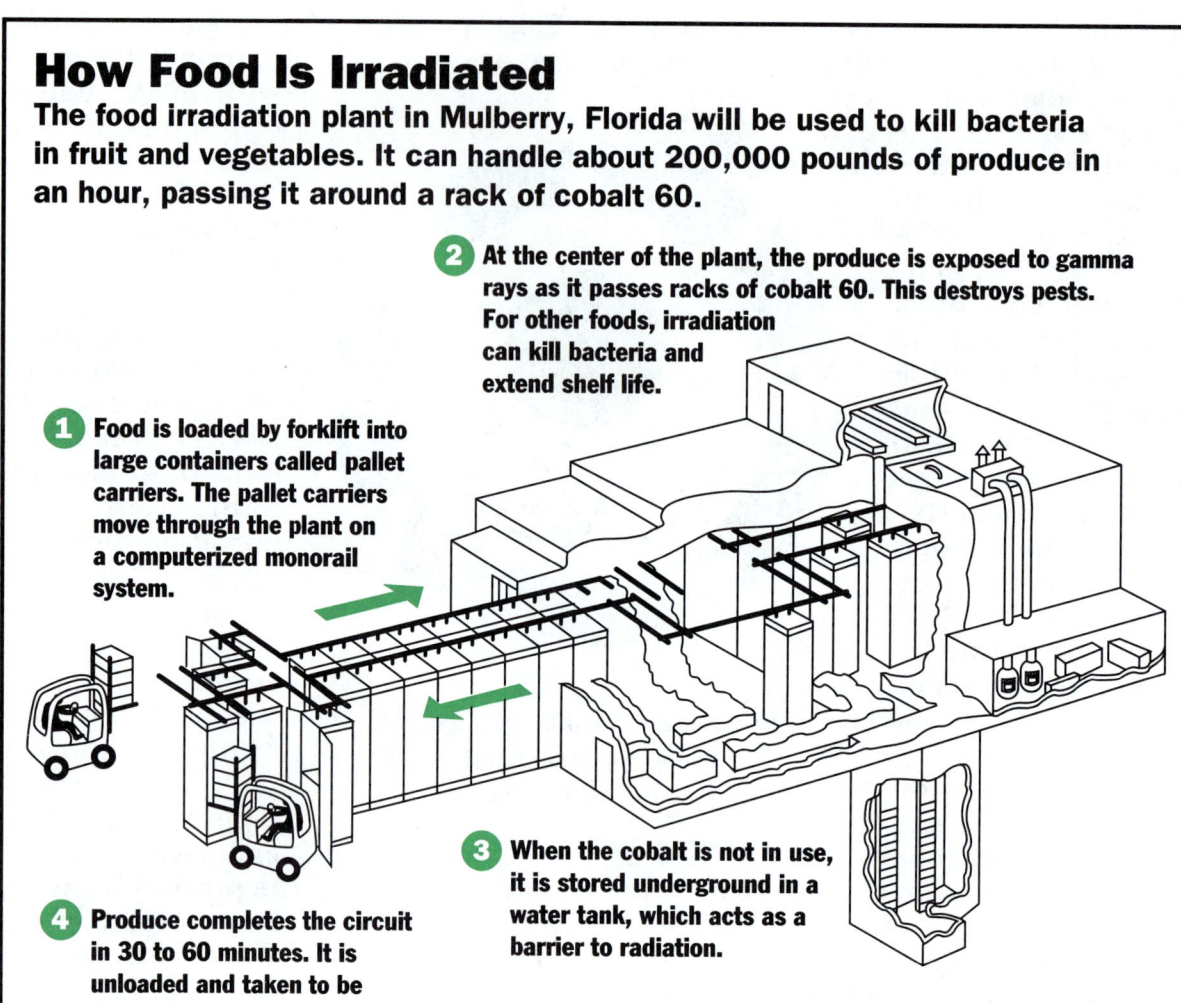

How Food Is Irradiated

The food irradiation plant in Mulberry, Florida will be used to kill bacteria in fruit and vegetables. It can handle about 200,000 pounds of produce in an hour, passing it around a rack of cobalt 60.

1 Food is loaded by forklift into large containers called pallet carriers. The pallet carriers move through the plant on a computerized monorail system.

2 At the center of the plant, the produce is exposed to gamma rays as it passes racks of cobalt 60. This destroys pests. For other foods, irradiation can kill bacteria and extend shelf life.

3 When the cobalt is not in use, it is stored underground in a water tank, which acts as a barrier to radiation.

4 Produce completes the circuit in 30 to 60 minutes. It is unloaded and taken to be processed.

Source: Nordion International
Adapted from: *The New York Times*

due in part to surveys showing widespread consumer doubts about the process, which industry officials say can be countered only with an expensive advertising and education campaign. The stores are also hesitant because of boycott threats by environmental and anti-nuclear groups, which mounted a radio advertisement campaign across Florida last summer in an effort of preventing the licensing of the Mulberry plant.

In efforts to win the public to their side, anti-irradiation organizations and consumer advocates have argued that while hundreds of scientific studies of irradiation have been conducted the results of many are inconclusive, ambiguous [unclear] at best or cautionary. They accuse the F.D.A. of acting hastily in response to pressures from growers and producers and ignoring negative studies that conclude that the process can have harmful effects.

"This is toxological Russian roulette," said Dr. Samuel S. Epstein, author of *The Politics of Cancer* and a professor of occupational and environmental medicine at the University of Illinois. "We just don't know where we stand, and until we do the necessary tests, it would be reckless to take an action that benefits only a few purveyors [suppliers] of marginal produce and a small industry looking for ways to dispose of nuclear waste."

Dr. Epstein's doubts about food irradiation have been echoed by a number of food safety and consumer advocacy groups whose leaders do not have scientific or medical training. But several respected scientists, including Dr. Louria, chairman of the Department of Preventive Medicine at the New Jersey Medical School in Newark, have expressed similar concerns about the lack of data and studies.

"Ethical and methodological barriers make it nearly impossible to study the effects of a diet of irradiated foods in human subjects," Dr. Louria wrote in *The Bulletin of the Atomic Scientists* in 1990. He also complained that three of the five studies cited by the F.D.A. "do not document the safety of food irradiation."

But an equally large, if not larger, group of experts, including those at food research institutes at major universities, maintain that research shows such fears to be unfounded. In a letter to the *Miami Herald* last month, Dr. Frank C. Lu, former chairman of the food safety program for the World Health Organization, said that the changes induced by food irradiation were also seen after storage and cooking of these foods.

"Irradiation actually reduces the number of harmful microorganisms, thus enhancing food's hygienic quality," Dr. Lu wrote.

After years of debate on the subject, legislatures in New York and New Jersey in 1989 enacted moratoriums [legal delays] on the sale or distribution of irradiated foods. New York recently extended its ban for two years, and New Jersey has also prohibited the "manufacture" of such items.

During the 1980's New Jersey emerged as the center of the nascent [newly born] irradiation industry as plants that primarily sterilized medical supplies located there. The sanctions by the New Jersey and New York governments came after residents and news organizations in those states expressed concerns that these plants would extend their activities to include fresh foods destined for consumption in the New York City metropolitan area.

Restrictions or outright bans on food irradiation exist in several countries, including Australia, Great Britain, Denmark, New Zealand and Sweden. Among countries that allow irradiated foods are China, France, the Netherlands and South Africa.

Though the Mulberry plant has been in operation since the beginning of the month, no fruits or vegetables have yet been shipped to grocery stores. Mr. Whitney said that some strawberries were irradiated last week, but that the berries' quality was not deemed high enough for them to be sent to market. . . .

Mr. Whitney said orders for the strawberries have also been received from stores in California, Pennsylvania, Rhode Island, and Washington [state], but he declined to name them for fear of alerting protesters. In addition, he said growers of other products have told him that they would adopt the process if

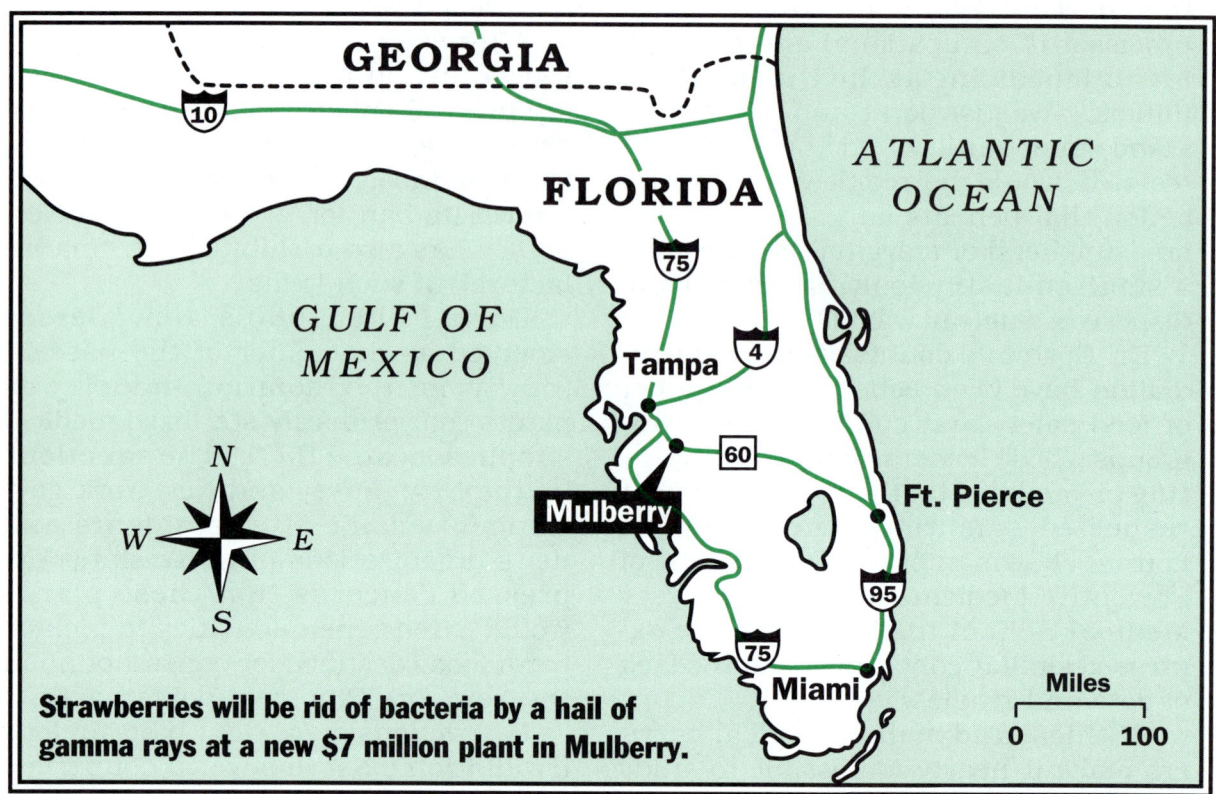

Strawberries will be rid of bacteria by a hail of gamma rays at a new $7 million plant in Mulberry.

Adapted from: The *New York Times*

there was not a public outcry over the strawberries.

Group With a Goal

. . . Federal regulations require all irradiated foods to display a circular, flower-like symbol, along with a label containing the phrase "Treated With Radiation," so consumers will know what they are buying. But Mr. Colby said the rules were "full of loopholes" that would allow irradiated strawberries, for instance, to be used in pies or jellies or sold in restaurants, hospitals and school cafeterias without that identification.

"You're really relying on the honesty of producers," he said, "and that's not enough."

Mr. Whitney, a trucking executive, dismisses such arguments as "sleazy lies" and cites numerous endorsements of irradiation from scientists at laboratories operated by universities, state departments of agriculture, the federal government, and industry groups.

The 3,000 residents of Mulberry appear to be unsure as to which side is correct. Though eager to attract new business that would lessen their dependence on phosphate mining, several residents and officials have expressed reservations about the irradiation plant, saying the 35 jobs it may offer do not compensate for increased safety risks.

"This thing has been rammed down our throats," said Andrew Scrocca, a town commissioner who led efforts first to prevent the plant from being built and then from opening. "We're being used as a test case, guinea pigs for an unproven technology. The government approved silicone breast implants, and now they're saying that's no good anymore. Who's to say this is any different?"

Fred Harris, safety officer at the Mulberry plant, says [owners of the plant and] the Canadian company that built the plant, have taken every precaution to prevent any equipment malfunctions or nuclear contamination.

But Michael Upledger, Florida program director for Food and Water, contends that even the most basic safety procedures at the plant were inadequate.

Mr. Scrocca said he feared that "the worst of the battle still lies ahead."

"If the consumer doesn't buy your product," he said, "you're out of business, and this is a town that can't afford that luxury."

5 GETTING INVOLVED

Gathering Information

Interview an Expert Invite a local chemistry or physics college professor or high school science teacher to your class. Be prepared to ask questions about radioactivity. For example, Are there different types of radioactivity? What are the major sources of radioactivity? What are the benefits and dangers of radioactivity? If you cannot interview an expert, go to your local library and research one of the questions mentioned above. To find information on radioactivity, use the card catalog. For some hints on using the card catalog, see page 238. Summarize your findings in a written report.

Cooperative Learning

Develop a Public Opinion Survey As a class, develop a public opinion survey to ask of area residents. Ask questions that concern that person's feelings on irradiated food. Would he or she be willing to purchase irradiated food? Does he or she think irradiated food is harmful? Does he or she feel there are any advantages to irradiated food? Disadvantages? Each person in the class should have two people respond to the survey. Analyze your results as a class.

Decision Making

Defend Your Decision Working in small groups, consider the arguments for and against food irradiation. Prepare a list of the arguments on both sides of this issue. Then, as a group, try to reach consensus either for or against it.

Next, go to your school or local library and find out who Florence Kelly was—why and how she formed the National Consumers League. Using this information, develop a plan that you would follow to ensure that your group's stand on the issue of food irradiation would be adopted in the United States. Present your plan to the class. Have the class evaluate how effective your proposed plan might be.

6 FOCUS ON...
Genetic Diversity

Today, much attention is being paid to the threatened extinction of plants and animals. But there is one aspect of this extinction that some scientists think is not getting enough attention. That is that we are putting our food supply in danger by the loss of genetic diversity in plants (or animals). Genetic diversity refers to the existence of a wide variety of different types of plants. Read the following article, and find out what is causing this slow, steady loss of plant diversity and why it can adversely affect our world food supply.

BEFORE YOU READ

Consider the headlines in the boxes below. How would these events affect humans? Think in terms of jobs, the economy, food prices, and the availability of foods.

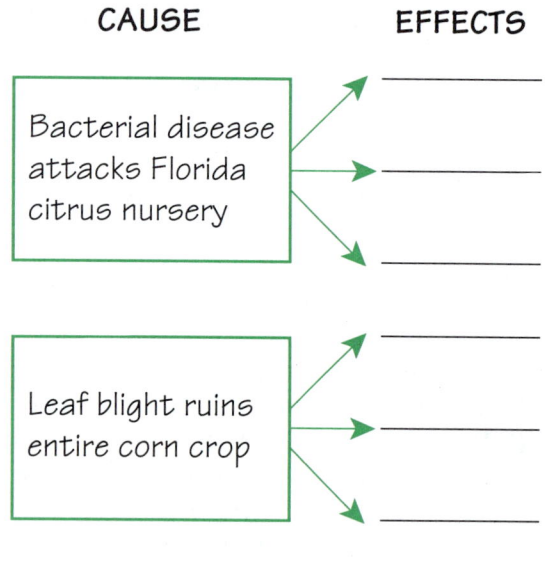

WHILE YOU READ

Copy the chart below onto a separate sheet of paper. While you read, pay close attention to the factors that cause scientists to want to increase research on genetic diversity. List each reason in the chart below.

Preserving genetic diversity in plants to safeguard against famine	
Crops we rely on for food	
Factors that lead to loss of genetic diversity	
Results of the loss	
Ways to preserve genetic diversity	

220

Genetic Diversity Prevents Blight, Spread of Famine

By Kitta MacPherson
FROM: *THE STAR LEDGER*—NEWARK, NEW JERSEY
JUNE 10, 1991

It may not sound like much to most, but when you mention "genetic diversity" to a group of plant geneticists [people who study genes], you can really get things humming.

That's because those who are experts in plant life are so worried about the issue that they believe the problem has reached crisis proportions.

Diversity in this case means just what it sounds like. Botanists and plant geneticists believe that there is value in the fact that the world is full of different varieties of plants, many of them unknown. Even within a certain type of plant, there are differences, such as wild varieties.

If the topic leaves you cold, think of this. There are precious few crops that we rely on for our food supply that are indigenous [native] to the United States. When you crunch into corn flakes, remember that corn was first grown by Indians in Mexico. Peanuts are from the fringes of the Amazon rain forest, Halloween pumpkins are from Central America, and potatoes are from Peru.

The only plants native to the United States that are now grown as crops are sunflowers, pecans, Jerusalem artichokes, cranberries, blueberries and wild rice.

The United States is not alone in its reliance on foreign seeds. Every country in the world has had to augment [increase] its supply of plants through the acquisition [obtaining] of seeds from elsewhere.

The topic is expected to get a full airing later this month in Washington, where a panel of scientists, headed by Peter Day, a world famous plant geneticist at Cook College of Rutgers University in New Brunswick [New Jersey], will examine "Genetic Diversity: Saving the World's Food Supply." Day, chairman of the National Research Council's committee on global genetic resources and director of Rutgers' Center for Agricultural Molecular Biology (Agbiotech), will be joined by other world leaders in the field in examining the South American plants that are essential to global agriculture.

The loss of plant species, many of them not yet collected or catalogued, is known as genetic erosion. Best estimates are that for some crops, we are losing 5 percent each year of crop diversity. This is due to a host of factors, including replacement by more marketable varieties, deforestation, overgrazing, water control projects, air pollution, and urbanization.

Why does this matter? Wholesale loss of irreplaceable plant diversity has the potential to destabilize [make unstable] the world food supply.

Widespread use of closely related varieties of a crop can set the stage for a new insect or disease to potentially wipe out the entire harvest, according to H. Garrison Wilkes, professor of biology at the University of Massachusetts. The world's developed nations are not immune to [free from] such disasters, and have come close to major devastation in recent years.

For example, in the United States, a new, little-known strain of leaf blight ruined 10 to 15 percent of the corn crop in 1970, and in 1984 a bacterial disease forced Florida nurseries to destroy 18 million citrus trees and seedlings.

These are early warning signs, experts believe, that must be taken seriously, especially for a growing world population. Experts predict an increase of one billion more people to feed within the next 10 years. One way to lessen the vulnerability [chances of attack] of the food supply is through plant breeding, which requires the use of naturally diverse plant species.

> "For the past 25 years, the driving force in food production has been to increase per-acre yields of food crops. This effort has been successful—with the doubling of the world population came the doubling of the food supply."

One innovation [something new] to be presented by John Dodds, head of the genetic resources department at the International Potato Center in Lima, Peru, is the "hairy potato." The potato resists most types of insect damage and requires no pesticide use. This is of vital interest to potato farmers who are among the heaviest users of insecticides in the world. In developing countries, such farmers spray an estimated $500 million worth of the products a year, much of that hazardous compounds that are banned in the United States and Europe.

The plant has earned the odd name because it protects itself by putting up a protective barrier of tiny hairs on the leaves of the plant. The barrier kills [many species of] insects on contact, but is believed to be harmless to humans and wildlife.

Research by another scientist points to the value of wild varieties of crops. Masaru Iwanaga, head of the genetic resources unit at the International Center for Tropical Agriculture in Colombia, has been breeding a wild bean plant found in Mexico over 20 years ago that repels bean weevils. These weevils decimate [destroy] about 13 to 25 percent of the bean crops in storage in South America and Africa every year.

While sprays are available that kill the weevil, the scientists believe the wild bean plant has a high degree of resistance to weevils never before found in cultivated [grown as crops] beans. In 1990, plant breeders at the center successfully transferred the weevil-resistant trait into common dry beans. Scientists hope the genetically altered beans will save farmers millions of dollars a year.

It all goes to a new theme in agriculture, promulgated [made known] by the Rockefeller Foundation in New York, known as "sustainability." For the past 25 years, the driving force in food production has been to increase per-acre yields of food crops. This effort has been successful—with the doubling of the world population came the doubling of the food supply.

The new shift changes the focus back to the long-term. High yields must be produced but without depleting the land irrevocably [making it impossible for the land to regain its fertility]. The environment should not be degraded [damaged], and farmers should be able to use affordable amounts of fertilizer and insecticides. Preserving genetic diversity is an essential part of this new world view.

6 GETTING INVOLVED

Decision Making

Library Research One attempt to prevent the loss of genetic diversity of plants is to form seed banks. Go to your local library and do research on what a seed bank is. What role have international agencies played in setting up seed banks? What role has the International Board for Plant Genetic Resources played in regard to seed banks? What happens to seed banks in times of political unrest in the countries where they exist? How successful have existing seed banks been? Do you think seed banks can solve the problem of loss of genetic diversity of plants? Prepare a report summarizing the advantages and disadvantages of seed banks.

Gathering Information

Read More About It The article pointed out that deforestation is one of the causes of loss of genetic diversity. Go to your school or local library and find out how much land is deforested each year. The *World Almanac* or an encyclopedia is a good place to start looking. Prepare a map that indicates where in the world serious deforestation is occurring. Display your map in the classroom.

Critical Thinking

Make the S–T–S Connection You know that science, technology, and society each affect one another. Think about the article you have just read. Express as clearly as you can the S-T-S interactions. Organize your ideas into a diagram, using arrows to show the flow of ideas between scientists, technology, and members of society.

7 FOCUS ON...
Genetically Engineered Crops

Most of you would probably agree that hazardous chemicals should not be used to protect crops from pests. But how are we going to protect crops from these pests while still meeting the growing need for a plentiful food supply? One answer to this problem comes from genetic engineering. This is the technological branch of genetics in which scientists directly manipulate genes for some desirable outcome. Read the following article to find out about the many ways scientists have manipulated genes to produce pest-resistant plants. Then decide how you feel about the controversial issue of releasing genetically engineered organisms into the environment for the benefit of agriculture.

BEFORE YOU READ

Look at the words in the chart below. Predict the meaning of each word. Discuss your predictions with a partner. Find out the actual meanings and compare responses.

WORD	Predicted Meaning	Actual Meaning
entomologist		
altered		
inoculate		
toxins		

WHILE YOU READ

This article discusses numerous facts about the bacterium Bt and its use in protecting plants from insect attack. Jot down some of the important facts about Bt in a chart like the one below. How can genetic engineering make use of Bt?

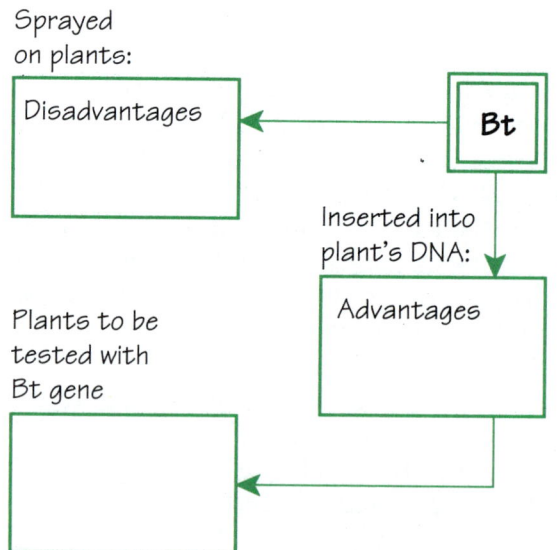

Scientists Use Gene Alterations to Make Crops Resistant to Infestation

By Judy Berlfein
FROM: *THE LOS ANGELES TIMES*
OCTOBER 14, 1991

Gary Reed has been spying on his potato plants. What he sees comes as a welcome surprise. In one 50-row section of his 1.6-acre plot, a number of beetles have become transfixed. "They're afraid to move, afraid to fly and afraid to eat," the Oregon State University entomologist [one who studies insects] said.

Adjoining this thriving growth, Reed has cultivated another area. Here the beetles have engaged in their normal activity—munching the green leaves and leaving behind only spindly stalks. The beetles have had a heyday [period of greatest success or fun] chomping away at the free lunch.

Both plots are free of chemical pesticides. The difference between the two: The vibrant green bundles have been genetically engineered to resist beetle infestation [harm caused by large numbers of insects].

Scientists have incorporated insecticidal [substance used to kill insects] genes into the plants' DNA. Taken from the bacterium *Bacillus thuringiensis,* the gene produces a protein that is toxic to beetles, caterpillars and flies, though harmless to humans and beneficial insects.

B. thuringiensis, commonly referred to as Bt, has been sprayed by farmers as a benign insecticide for 30 years. When attacking bugs gnaw at the plant leaves, they also ingest [swallow] a protein called Bt toxin. The protein wreaks havoc on [causes great disorder to] the insects' digestive systems. The little critters stop eating and die of starvation.

But because Bt is an expensive pesticide, it can be used only on high-value crops. Furthermore, the Bt toxin is degraded [broken down into simpler substances] rapidly in the environment and washed off by rainfall.

For these reasons, botanists would like to incorporate it into the plant, reducing the cost of use and providing permanent protection throughout the lifetime of the plant. That is precisely what Reed has achieved in his potato plots.

Plant biology is undergoing a revolution more profound than any since the monk Gregor Mendel discovered the principles of the inheritance of genes

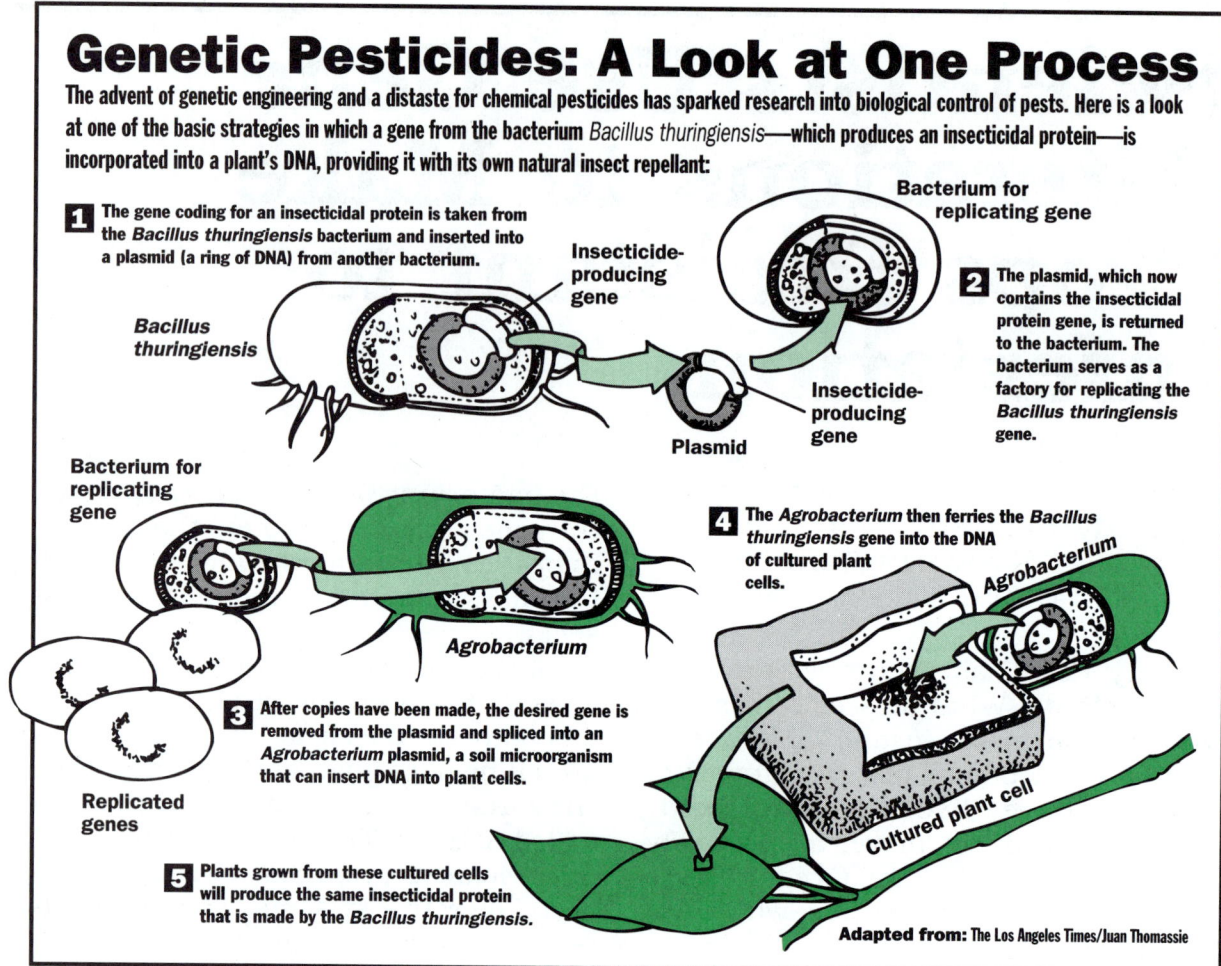

Genetic Pesticides: A Look at One Process

The advent of genetic engineering and a distaste for chemical pesticides has sparked research into biological control of pests. Here is a look at one of the basic strategies in which a gene from the bacterium *Bacillus thuringiensis*—which produces an insecticidal protein—is incorporated into a plant's DNA, providing it with its own natural insect repellant:

1 The gene coding for an insecticidal protein is taken from the *Bacillus thuringiensis* bacterium and inserted into a plasmid (a ring of DNA) from another bacterium.

2 The plasmid, which now contains the insecticidal protein gene, is returned to the bacterium. The bacterium serves as a factory for replicating the *Bacillus thuringiensis* gene.

3 After copies have been made, the desired gene is removed from the plasmid and spliced into an *Agrobacterium* plasmid, a soil microorganism that can insert DNA into plant cells.

4 The *Agrobacterium* then ferries the *Bacillus thuringiensis* gene into the DNA of cultured plant cells.

5 Plants grown from these cultured cells will produce the same insecticidal protein that is made by the *Bacillus thuringiensis*.

Adapted from: The Los Angeles Times/Juan Thomassie

---

125 years ago. Molecular biologists are modifying agricultural crops to give them a variety of new characteristics, including resistance to pests and herbicides, tolerance to salt and drought, and increased nutritional properties. . . .

But one of the principal focuses has been on providing increased resistance to insects, because of the public's growing intolerance [unwillingness to endure] of the use of chemical pesticides and the mammoth size of the pesticide industry—$1 billion per year in the United States and $5 billion worldwide. The biotechnology companies hope to grab a significant share of that market, enticing farmers to buy seeds that might cost 50% more than normal, but that would save several times the extra cost in reduced use of pesticides. Much of that research focuses on Bt. . . .

But years of work still lay ahead. Tobacco had been used only as a model. It was plants like potatoes, tomatoes and corn that would benefit from the Bt toxin—plants that suffered specifically from caterpillar and beetle damage.

Since then, numerous companies have succeeded in transferring the Bt toxin gene into the desired tomato, cotton and corn plants. The plants stave off [keep off] predators without harming beneficial insects in the process. They also seem to produce no adverse environmental effects. Industry analysts predict that the first genetically engineered crops, such as cotton, could be approved by the government by 1995.

Some companies have tried to circumvent [go around] the cumbersome [long] regulatory process by using genetic engineering techniques on the pesticide rather than the plant itself and are therefore more likely to win speedy approval. . . .

While enthusiasm for Bt is strong, none of the developers imagine that biological insect control will ever put the chemical companies out of business. "There have been claims that genetic engineering is going to let us eliminate chemical pesticides," [Ron] Muessen [a director of plant biotechnology research] said. "That's not going to happen. The number and complexity of pest problems that growers face is sufficiently daunting that we're not going to completely abandon one set of tools."

7 GETTING INVOLVED

Gathering Information

Library Research The article mentions that some companies are trying to avoid the regulatory process involved with genetic engineering. Go to your local library and find out if genetically engineered foods must be approved by the Federal Food and Drug Administration. Also, find out if special labels are required for these foods. You may be able to find the information using a newspaper index. For some tips on using a newspaper index, see page 236. Prepare a brief summary of your findings.

Decision Making

What Do You Think? Answer the following questions in complete sentences. Remember, there are no right or wrong answers—these are your belief statements.

1. Should foods that have been genetically altered be labeled for consumer protection?
2. Would you prefer to eat genetically altered food or foods treated with pesticides?
3. Do you think the process of altering plant genes is a good method of controlling pests?
4. What could be a possible danger of genetic engineering?

Cooperative Learning

Read More About It Find out who Gregor Mendel was and what his experiments proved. What principles of inheritance did Mendel discover? Why were his experiments so important? What questions about heredity did his research answer? Work in small groups to prepare an exhibit display that summarizes what you have learned about Mendel. Display your exhibit in your classroom.

World Food Resources

THINKING CRITICALLY ABOUT LOCAL ISSUES

You may want to work with a group of students for this activity. If so, make sure that each member of the group agrees about the issue that is chosen.

Prioritize the Issues Make a list of food resource issues that are important to you and your community. Start by reviewing the brainstorming list that you made at the beginning of this module. If you make a scrapbook, or wrote in a journal, look over those as well. Also, review the magazine and newspaper articles in this module. Evaluate all the issues on your list. Then, put them in order so the issue that has the highest priority for you or your group is at the top of the list.

Make the Issue Your Own Become an expert on your issue. Then decide what you as a citizen can do about the issue. Apply the process that you learned in the *S-T-S Problem Solving* module in the front of this book. Remember, the skills in the process are:

- Analyzing the Issue
- Gathering Information
- Making a Decision
- Planning Action

If you are working in a group, each group member has to contribute to all four stages. After completing the process, share your experience with your class.

CREATIVE THINKING

Design and write a 30-second television advertisement that would increase awareness of world hunger and motivate people to take action. Before you begin, think about what you want the central message to be and what you would like as a response from the audience. Include a complete description of the audio and video portions of the announcement. If video equipment is available, you might consider trying to record your advertisement.

Wrap-Up

HELP WANTED

State a Panel Discussion Suppose a panel of experts were assembled to discuss possible solutions to world hunger. The experts included a food technologist, a grain farmer, an agricultural engineer, a nutritionist, a land-use planner, and a political scientist. Find out about each of these careers and explain what kind of information each person would bring to this issue.

As a class, stage a discussion on the issue. One student should act as the moderator while six other students should play the roles of the experts listed above. The rest of the class should play the roles of interested citizens asking questions of the panel of experts.

One way to inform people of the problem of world hunger is to make a commercial.

SKILLS AND RESOURCES

CONTENTS

Skills for Gathering Information
Brainstorming	231
Making a Scrapbook	232
Keeping a Journal	233
Writing a Letter to Request Information	234
Using the Telephone Effectively	235
Organizing Data in a Data Table	236
Using the Catalog in the Library	238

Skills for Taking Action
Writing a News Release	239
Making a Persuasive Speech	240
Writing a Letter to a Member of Congress	242
Setting Up an Awareness Fair	243

Community Resource Directory
General Organizations	244
Organizations for Young Adults	252
Publications of Interest	253
Government Resources	256
Science Museums and Zoos	262
Career and Trade Organizations	266

Skills for Gathering Information

BRAINSTORMING

What Is Brainstorming?

Brainstorming is a cooperative technique used to collect many ideas on a particular topic in a short amount of time. Brainstorming involves a group of people who rapidly throw out ideas without stopping to evaluate each one.

How to Brainstorm

1. *Decide on who will be involved in the brainstorming session and where the session will be held.* A brainstorming group should consist of at least three people. Do not make the group too large; it is difficult to keep a large group focused.
2. *Choose one group member to act as a facilitator.* The facilitator has the responsibility of making sure the session runs smoothly. He or she should make sure that every member of the group has a chance to speak and that the group stays focused. If anyone has not contributed, the facilitator should encourage him or her to do so.
3. *Choose another group member to act as a recorder.* The recorder is responsible for writing down all the ideas generated during the brainstorming session.
4. *Clearly state the topic of the brainstorming session.* Write the topic on the chalkboard or a large sheet of paper where everyone can see it.
5. *Give all group members a chance to offer their ideas.* For a more orderly session, group members should raise their hands and wait to be called on by the facilitator.
6. *Listen closely to the ideas offered by others.* Often one person's ideas will cause others to think of additional ideas.
7. *Do not criticize any of the ideas.* Remember, the goal of a brainstorming session is only to think of ideas: they should be evaluated at a later time.
8. *Be creative.* Feel free to offer all of the ideas that come to mind. The best ideas often seem a little bit crazy at first.

Thinking About the Skill

Why is it important that criticism of ideas not be allowed during a brainstorming session?

MAKING A SCRAPBOOK

What Is a Scrapbook?

A scrapbook is an organized collection of newspaper articles, magazine articles, pamphlets, photographs, and other items that are related to a particular topic. Creating a scrapbook actively involves you in a search for evidence and information about a topic. Once it is created, a scrapbook can serve as a reference and reminder of what you learned about the topic.

How to Make a Scrapbook

1. *Identify the purpose of your scrapbook.* For example, the purpose may be to focus on how an issue affects your town. You could either explore one aspect of a topic in depth or collect as much information as possible on the whole topic.
2. *Decide on what kinds of items to collect.* To fulfill the purpose of your scrapbook, you may decide to collect newspaper articles, pamphlets, photographs, advertisements, and other similar items. Sometimes your teacher may instruct you to collect only one of these types of items.
3. *Gather the items for your scrapbook.* This may involve telephoning or writing to appropriate organizations or looking through newspapers and magazines. Make sure you get permission to cut up or take newspapers, magazines, and other items that do not belong to you.
4. *Choose a notebook in which to display the items.* Make sure that the notebook is large enough to hold all the items that you collect. You can create your own notebook by stapling together sheets of blank paper and adding a cover. Add drawings or other decorations to make it attractive. Large sketch books also make good scrapbooks. These can be found at most art supply or variety stores.
5. *Attach the items to the pages of your notebook.* Arrange the items in a logical order before attaching them to the scrapbook. Use glue or tape for permanent attachment. Use paper clips or staples for temporary attachment. Date or label each item in your scrapbook.
6. *Create a title page for your scrapbook.* Include a title, the date, your name, and any other kind of information that you feel is important and worthwhile.

Thinking About the Skill

Suppose you wanted to create a scrapbook about plants and animals that are threatened with extinction. What kinds of items would you collect? Describe two ways you could organize the items in your scrapbook.

KEEPING A JOURNAL

What Is a Journal?

A journal is a place where you can keep a personal record of events or a log of your daily activities. You can keep a journal on a specific topic, or use it to record a variety of daily events. Journal writing provides the opportunity to record your thoughts about *topics* about which you are learning and thinking. Unlike a diary, which is about *you*, journals are often shared with others.

How to Keep a Journal

1. *Choose a single notebook in which to record all your journal entries.* Make sure that your notebook has enough pages to accommodate journal entries over a long enough period of time.
2. *Identify the topic of your entry.* It may be a topic of your own choosing or one assigned by your teacher. Write down the topic and date of your entry.
3. *Write your thoughts about the topic.* Use your journal to make sense of the topic for yourself. If you are logging your activities, use a list form as illustrated. Otherwise, use a narrative form, also illustrated below. An entry may be of any length, but it should be long enough to express what you want to record.
4. *Be ready to share your journal with others.* You may want to exchange journals with another student as a way to share your thoughts. You may also want to share your journal with your teacher. Journal entries can also be shared during a group discussion with other students who have written about the same topic.
5. *Write in your journal often.* You can use your journal to evaluate certain things you do in order to improve on them. For example, if you find that you are eating a lot of junk foods, you can use your journal as a place to record ideas about how to change your behavior. You can also use a journal to work on drafts of reports, and to keep a record of your work to date.

List Form

	Foods That I Eat	Amount
morning	doughnut	1
	whole milk	12 ounces
afternoon	bologna	3 slices
	white bread	2 slices
	potato chips	2 ounces
	fruit punch	12 ounces
evening	chicken	1/4
	peas	1/2 cup
	mashed potatoes	1 cup
	milk	12 ounces

Narrative Form

November 12: My Ideas About the Foods I Eat

For several days I've been writing down the foods I eat. I think there are some changes I can make in my diet that would bring it more in line with the new food pyramid. For example, I should eat more fresh fruits and vegetables. Cereals with fruit would be a better breakfast than a doughnut.

Thinking About the Skill

Imagine that you kept a journal of your thoughts about foods that you have seen thrown away in your home, school, and town. How might you use this information in the future?

WRITING A LETTER TO REQUEST INFORMATION

What Is a Letter to Request Information?

A letter to request information is a letter that you send to a person or organization, asking for specific information. You use this method to obtain information that is not available from other sources.

How to Write a Letter Requesting Information

1. *Find out the names and addresses of sources that are likely to have the information you need.* Often, the title of an organization or agency indicates the kind of information that it can provide.
2. *Type or write your letter so it is easy to read.* If your letter is difficult to read, chances are it will not be answered. Include your return address and signature.
3. *Be specific in your request.* Don't ask for everything. Specify only the information you need. This way, the person who answers your request is more apt to send you what you want.

> 456 E. 19th Street
> New York, NY 10003
> January 6, 1993
>
> OXFAM
> 115 Broadway
> Boston, MA 02116
>
> Dear Sir or Ms.:
>
> As part of a school project, I am writing a report on hunger and war. Would you please send me any information that you have about hunger in countries that are at war? Thank you for your help in this matter.
>
> Sincerely,
>
> *Ronald Moore*
>
> Ronald Moore

If a typewriter is available, you should use it when you write letters.

4. *Be polite in your request.* Remember, you are asking for information, not demanding it. A courteous request is more productive than one that is not. Use phrases like "please send me" rather than "send me."
5. *Use the proper form for a business letter.* Start the letter with "Dear Sir or Ms.:" and end the letter with "Sincerely," followed by your signature and your name, typed or printed neatly.
6. *Proofread your letter for mistakes.* Check the address, spelling, and grammar.
7. *Mail your letter.* If you have followed these guidelines, chances are you will receive a reply.

Thinking About the Skill

Sometimes, in busy offices, letters requesting information get set aside. What actions could you take if you do not get a reply in a few weeks?

USING THE TELEPHONE EFFECTIVELY

Why Use the Telephone to Gather Information?

Using the telephone can be the quickest and most convenient way to gather certain kinds of information. Sometimes, it may be the only way.

How to Use the Telephone to Gather Information

1. *Identify the exact information you want.*
2. *Identify the source(s) that are likely to have the information you want.* The sources may be government or private agencies, companies, institutions, public officials, or individual experts.
3. *Find the telephone number(s) of the source(s) you want to call.* Refer to a telephone directory for the city or region where the person or organization is located. If you do not have access to a telephone directory, call the information operator for the area you want to reach. For telephone numbers within your area code, call 411. For numbers outside your area code, call 1–(area code)–555–1212. You can find the area codes in the front pages of any telephone directory.
4. *Prepare a telephone data sheet.* This is a sheet of paper on which you keep a record of your calls and any information that you may need while you are on the telephone. Write a list of questions you want to ask and the name and telephone number of the person or organization you are calling. Leave space to write the information that the person may provide.
5. *Get permission to use the telephone.* This is especially important if you need to make long-distance telephone calls.
6. *Make the telephone call during regular business hours.* If you are calling an organization in another region of the country, be sure it is between 9AM and 5PM at the location you are calling.
7. *Identify yourself, giving your name, school, and grade.* If you have the name of a person, ask to speak to that person; if you do not, ask for a person who is in a position to know the information you are seeking. If the person you need to speak to is not available, find out when he or she will be available and call back then. Otherwise, leave your name, the reason for your call, a telephone number, and the time at which you can be reached.
8. *Record on your data sheet the information the person provides.* Since the person is likely to speak faster than you can write, politely ask him or her to repeat the information you missed. Record the name, title, address, and phone extension of the person who gives you the information. This could save you time and effort if you need to contact the person again. It also allows you to cite the person as the source of the information.
9. *Thank the person for giving you the information.*

Thinking About the Skill

List the advantages of gathering information by telephone rather than by writing a letter.

ORGANIZING DATA IN A DATA TABLE

What Is a Data Table?

A data table is a chart consisting of horizontal rows and vertical columns in which data are systematically arranged. You make a data table to see what patterns or relationships may exist among a large amount of data.

How to Make a Data Table

1. ***Collect your data.*** Data are generated by experiments and surveys. Suppose, for example, you conducted a survey at your school. In the survey, you asked students if they approved or disapproved of requiring health-care workers to be tested for AIDS. On their survey forms, students also indicated their grade level and whether they were male or female. The raw, or unorganized, data from the survey are as follows:

2. ***Decide how to organize the data in a table.*** A data table should show all the data in a way that allows relationships or patterns to be seen. In the case of your survey, for example, you would want to arrange the data so the opinions of students were grouped, or categorized, according to both the students' grade levels and their sex.

disapprove	grade 9	female
disapprove	grade 7	male
approve	grade 7	male
approve	grade 8	female
approve	grade 8	female
disapprove	grade 9	male
disapprove	grade 7	male
approve	grade 9	male
disapprove	grade 7	male
approve	grade 7	female
approve	grade 7	male
approve	grade 8	female
approve	grade 9	female
disapprove	grade 9	male
approve	grade 8	female
approve	grade 9	female
disapprove	grade 7	male
disapprove	grade 8	female
approve	grade 7	male
disapprove	grade 9	male
disapprove	grade 7	male
approve	grade 7	female
approve	grade 8	female
approve	grade 7	female
approve	grade 9	female
disapprove	grade 8	male
approve	grade 9	female
approve	grade 7	male
disapprove	grade 8	male
approve	grade 9	female
approve	grade 8	female
disapprove	grade 9	male
disapprove	grade 7	male
disapprove	grade 9	male
disapprove	grade 7	male
approve	grade 7	female
disapprove	grade 8	female
disapprove	grade 9	female
disapprove	grade 9	male
approve	grade 8	female
approve	grade 9	female
approve	grade 7	male
disapprove	grade 8	female
disapprove	grade 7	male
disapprove	grade 9	male
disapprove	grade 7	male
approve	grade 7	female
disapprove	grade 8	female
approve	grade 8	female
approve	grade 9	female
disapprove	grade 8	male
approve	grade 9	female
approve	grade 7	male
approve	grade 8	male
disapprove	grade 8	female
approve	grade 8	female
approve	grade 9	female
disapprove	grade 8	male
approve	grade 9	female
disapprove	grade 7	male
disapprove	grade 8	male

Number of Students Who Approved and Disapproved of Required AIDS Testing for Health-Care Workers

	Grade 7	Grade 8	Grade 9	Male	Female
Approve					
Disapprove					

3. ***Draw a rough layout of the table.*** This step quickly shows how the table will look and if it needs to be changed or modified. Decide what the categories of data will be and how they can be arranged in rows and columns. As you can see in the sample table, the five student categories, "Grade 7," "Grade 8," "Grade 9," "Male," and "Female," can be used as the headings of the columns. The two opinion categories, "approve" and "disapprove," can be used as the row headings.

4. ***Draw the table.*** Using a straight edge, draw vertical lines to form columns and horizontal lines to form rows. Write the column and row headings clearly.

5. ***Give your table a title.*** The title should define the information in the table, including what the numbers represent. For example, "Number of Students Who Approved and Disapproved of Required AIDS Testing for Health Care Workers."

Thinking About the Skill

Look at the data in the sample data table above. What patterns or relationships do you see?

USING THE CATALOG IN THE LIBRARY

What Is the Card Catalog?

The card catalog is an alphabetical file, usually on 3 × 5 cards, of all the books in the library. Many libraries now have computerized catalogs rather than, or in addition to, card catalogs. Instructions for using computerized catalogs are displayed on the computer screen and posted beside each computer workstation.

How to Use the Card Catalog

1. *Identify the book's title or author or the subject area in which you are interested.* You can search in either a card catalog or a computerized catalog by using one of these three categories.

2. *Search alphabetically in the catalog for the book's title, the author's name, or the subject area.* In a card catalog, each book's identification number and basic information are recorded on a 3 × 5 card. However, for each book, three different cards are filed—a *title card,* with the book's title at the top, an *author card,* with the author's name at the top, and a *subject card,* with the subject area of the book at the top. The cards are filed alphabetically in a set of labeled drawers. Some libraries have one card catalog with title, author, and subject cards filed together alphabetically. Other libraries have two catalogs, one just for authors, the other for title and subject cards.

   To search for a book by title, ignore first words such as *A, An,* or *The.* For example, the book *The Closing Circle* is filed alphabetically under the C's.

   To search for a book by author, look for the last name first as you would look for a person's name in the telephone directory. For example, for a book by Vicki Cobb, search under Cobb, Vicki.

   To search for a book by subject matter, search alphabetically under that subject word. From among all the books under a particular subject, you would select the one that appears to have the information that you need.

   If you use a computerized catalog instead of the card catalog, you can search in any of the categories at the same time by typing in a name or word. The computer helps you to complete your book search step-by-step.

3. *Write down the title, author, and call number of the book you want.* The call number is at the upper left corner of the card.

4. *Search the section of the library where your book should be located.* Refer to the call number labels at the ends of each row of shelves. Find the book by looking at the call number on its spine.

5. *If the book you want is not on the shelf, consult the librarian.* The librarian can find out if the book is checked out or in a different location. If it is, you may want to look at other books with similar call numbers.

Thinking About the Skill

How could you use the catalog to help you find all the books on a particular topic in a library?

Skills for Taking Action

WRITING A NEWS RELEASE

What Is a News Release?

A news release is an announcement about a product or an event that is prepared for release to the news media—newspapers, radio, television, and news magazines. You write a news release so that someone in the media can, in turn, communicate this information to a wide audience.

How to Write a News Release

1. ***Gather the names and addresses of the reporters to whom you will send your release.*** Plan to send your release to all media (publications, radio stations, television stations, and so on) with audiences that you want to reach. Whenever possible, address your release to a specific person. If you cannot find out the name of the particular reporter to whom to send the release, address your news release to the "News Editor."

2. ***Write a heading for your news release.*** At the top of a sheet of paper, clearly write or type the words "NEWS RELEASE." Below this, give the date when the information can be made public (usually, "For Immediate Release"). Also, give the name and telephone number of the person to contact for more information (it may be you). Finally, include the date on which the release is written.

3. ***Write the body of the news release.*** The body of your news release should answer "the five W's"—who, what, where, when, and why. Use short sentences and active verbs. If possible, include something that will attract a reporter's attention (for example, a humorous line, a photo, an intriguing question). Remember that editors receive many news releases so it is important to make yours stand out.

4. ***Keep your news release brief.*** It is best to type your release, double-spaced, and to keep it to one page. If you must continue onto a second page, type the word "MORE" at the bottom of the first page and include a page number at the top of the next page. At the end of the release, type "-30-." This is a code that means "the end."

5. ***Proofread your release.*** Check to see that your spelling and grammar are correct. Also, make sure that your statements are clear and factual and that you haven't forgotten any important details.

6. ***Mail or deliver a copy of your news release to each reporter on your list.*** Make sure to send out the release far enough in advance of the event (2–3 weeks) so that reporters have enough time to act on it.

Thinking About the Skill

Suppose you wanted to alert the people who live or shop in your town to an upcoming event at your school. List all the news media to whom you would send a news release.

MAKING A PERSUASIVE SPEECH

What Is a Persuasive Speech?

A persuasive speech is a speech that tries to influence the attitudes, beliefs, or behavior of an audience. You make a persuasive speech in order to change people's minds or move them to take some action.

How to Make a Persuasive Speech

1. *Decide on the purpose or goal of your speech.* For example, do you want to persuade people to change their views on an issue? Or do you want to convince people to take a specific action? Whatever your goal, make sure it is something that you truly believe in. If you yourself are not convinced of your goal, you cannot expect to influence others.

2. *Know your audience.* Do your listeners share your views on the issue or do they hold an opposing point of view? Or are they uncommitted, that is, they have not yet made up their minds? Or do they not care one way or another about the topic? Knowing your audience will help you focus your speech to meet their particular needs. But, what if, as in most cases, an audience includes a mixture of people with different attitudes? Then, gear your speech to the largest group in the audience.

3. *Write your speech.* Your speech should focus clearly on your goal and present enough evidence to back up your view. To best accomplish this, follow these pointers.

    a. **Use an attention-getting introduction.** A humorous story, a famous quotation, a startling statistic—these are some ways to grab your audience's attention. Listeners will be more interested in what you have to say if you get their attention right away.

    b. **Establish your credibility.** It is essential to give your audience a reason to believe you. To do this, tell them why you consider yourself an expert on your subject. You may, for example, have studied your subject for years or have had relevant personal experiences. In addition to discussing your expertise, ''name drop'' the names of well-known experts who share your view.

    c. **Present valid evidence and use clear reasoning to draw conclusions.** Build a case for your viewpoint by presenting supporting facts and examples. Also, present the research and opinions of scientists or other experts who share your view. Remember that your goal is to convince listeners that your viewpoint is the most logical one, given the evidence. Do not let your emotions overshadow the evidence.

4. *Practice your speech.* Practice your speech by yourself at first. Do not try to memorize the entire speech. Instead, use index cards to jot down a few words to remind you of each sentence or paragraph. That way, your delivery will be more natural. After you feel comfortable with your speech, practice it in front of friends and family members.

5. *Deliver your speech.* As the time approaches to deliver your speech, expect to feel nervous. One trick that helps control nerves is to pre-

tend that the audience is made up of family and friends. Once you begin to speak, you will probably relax and feel more confident. Other things to keep in mind while speaking are to talk slowly and loudly and to make eye contact with members of the audience. Do your best to come across as a sincere, likable person with a sense of humor and respect for your audience. Remember, it is easier to persuade people if they like you. The most important thing you have to sell is yourself.

> Students make own food choices
>
> We want to create healthy habits
>
> Want nutritional information in the school cafeteria
>
> Then decisions can be made

6. *Evaluate your performance.* After your speech is over, rate yourself on how well you did. Did the audience seem interested? Did some people seem persuaded by your views? How could you improve your performance? Jot down some notes to help you in your next speech.

Thinking About the Skill

Imagine that you are to give a speech on the need to protect endangered species. Your audience does not care at all about the issue. How would you get them to care?

It is helpful to practice a speech before a small group of classmates before speaking to a large public audience.

WRITING A LETTER TO A MEMBER OF CONGRESS

Why Write a Letter to a Member of Congress?

Citizens write to their representatives in Congress for a variety of reasons. Some simply ask for information, while others may request help to solve a problem they are having with a government agency. Most people who write, however, do so to express concerns or opinions about issues that will be voted on by Congress. In this way, you can influence the way your representatives will vote on an issue.

How to Write a Letter to a Member of Congress

1. *Identify the issue about which you are writing.* If the issue concerns a proposed law, identify the bill by its number, for example, H.R. 112, or by its popular title, such as the "Child-Care Bill." Often, newspaper and magazine articles about an issue mention the number or name of a bill that is being considered in Congress.
2. *Write about an issue while it is current.* Do not wait until a bill on an issue has already been voted on. Check newspapers for updates on the status of the issue.
3. *Use the proper form for your letter.* Begin your letter with "Dear Representative _____," or "Dear Senator _____," and end the letter with "Sincerely," followed by your name, grade, and school or organization. Also, be sure to include your return address.
4. *Express your opinion on the issue and give reasons for your opinion.* State the facts regarding the issue as you know them. Then, base the reasoning for your opinion on those facts. Remember, an opinion without facts and reasoning to back it up will not carry as much weight.
5. *Be brief and to the point.* Say only what you need to say. The most effective letters to Members of Congress are short and simple.
6. *Address your letter as follows:*

   Representative _____
   House Office Building
   Washington, DC 20515
   or
   Senator _____
   Senate Office Building
   Washington, DC 20510

7. *Reread your letter for its content.* Check to see if the letter clearly expresses your opinion and the reasons for your opinion. Have someone else read your letter to see if it is clear. Make adjustments to the letter as necessary.
8. *Proofread your letter for mistakes.* Take the time to correct any spelling or grammatical errors. If possible, type your letter. A letter that is easy to read will get more attention than one with hard-to-read handwriting.
9. *Wait for a reply.* In most cases, a member of Congress will write a letter in reply. Be patient. It may take several weeks for a response to finally arrive.

Thinking About the Skill

Why do you think writing to a Senator or a Representative is an effective way to influence the way he or she may vote on a particular issue?

SETTING UP AN AWARENESS FAIR

What Is an Awareness Fair?

An awareness fair is an event at which a variety of exhibits, demonstrations, discussions, activities, and talks are held to heighten people's awareness of a particular issue. Awareness fairs provide the opportunity for people to become informed about an issue in a social setting.

How to Set Up an Awareness Fair

1. *Identify an issue that you and others feel people should be aware of.* The issue can be any issue of current concern, for example, rain forest destruction, homeless people, or AIDS.
2. *Prepare a list of potential awareness projects.* Develop ideas for exhibits, demonstrations, and other activities that would foster awareness of the issue. Select ideas for projects that you and your classmates can realistically carry out.
3. *Assign each project to an interested person or group of people.* Each activity should be well-defined so students know clearly what they are doing and do not duplicate one another's efforts.
4. *Invite local organizations that are involved with the issue to participate in the fair.* Organizations that work on issues usually welcome opportunities to inform people about their work.
5. *Invite experts to present talks or lead discussions.* Local experts can be valuable resources for information on various issues.
6. *Select a time and place for the fair.* Consider the space, equipment, utilities, and schedule requirements of the participants. Be sure the location you choose is suitable for all the participants' needs.
7. *Advertise.* Place notices in your school and local newspapers and on community bulletin boards. Make posters and display them in your school and community.
8. *Monitor the preparations of all the participants.* Check with the participants on a regular basis to keep all the projects on schedule.
9. *Prepare a program.* The program should be a printed guide for anyone who attends the fair. It should show the time and location of all the fair's exhibits and activities.

These students are attending a health fair sponsored by a hospital.

Thinking About the Skill

What advantage does an awareness fair have over other methods of increasing people's awareness of an issue, such as letters to the editor?

Community Resource Directory

Are you interested in learning more about the four module topics discussed in this book? Do you have any questions that weren't answered? Do you want to get more involved in issues concerning extinction of living things, human population, human health and disease, and world food resources? Here are some ideas about how and where you can find out more.

GENERAL ORGANIZATIONS

Here's a list of helpful, nonprofit organizations from across the country. You may want to contact any or all of them to learn more about the four module topics discussed in this book. Many of these organizations have people on staff who can answer your questions. Others will send you written materials (brochures, newsletters, booklets, and so on) about matters discussed in this book. Still others offer ways in which you can get involved. Some do all these things.

You can either write to the addresses or call the telephone numbers given below. Asking for the public information, human resources, or education departments will usually get you the person you want. Remember that while calls to 800 numbers are free, you will be charged for other long distance calls.

Those organizations that focus only on limited issues may be evident by their name. Unless otherwise indicated, the other organizations provide information on a variety of related topics.

If you don't see an organization here that can answer your questions, keep in mind that there is an organization for almost every subject. Many of the organizations listed here can refer you to others that may be better able to help you. If there is a specific disease or topic you would like to know more about, try looking in the yellow pages in the telephone directory under its name. Furthermore, if you know of an organization that is not listed here, you can call the toll-free operator at (800) 555-1212 to see if there's a toll-free number for it.

American Association of Retired Persons (AARP)
[National Headquarters]
601 E Street NW
Washington, DC 20049
(202) 434-2277

This is a research center with information on aging and health. You can also call your state or local chapter of the AARP.

American Association for the Advancement of Science (AAAS)
1333 H Street, NW
Washington, DC 20005
(202) 326-6400

The AAAS is one of the oldest, largest, and most influential scientific organizations. It sends written materials but doesn't answer questions over the phone.

American Association for World Health
2001 S Street, NW, Suite 530
Washington, DC 20009
(202) 265-0286

The association offers a membership discount to students.

American Association of Zoological Parks & Aquariums
4550 Montgomery Avenue
Suite 940N
Bethesda, MD 20814
(301) 907-7777

Information is available on the roles that zoos and aquariums play in the protection of wildlife and fish.

American Cancer Society
[National Headquarters]
1599 Clifton Road NE
Atlanta, GA 30329-4251
(800) 227-2345 [800-ACS-2345]

Ask for the public education department and you will be directed to someone who can answer your question and/or send you requested literature. He or she will also tell you where the branch in your state is located.

American Civil Liberties Union (ACLU) [National Office]
132 West 43rd Street
New York, NY 10036
(212) 944-9800

The ACLU will send out formal statements on its position on immigration and other issues involving individual rights.

American Federation of Teachers
555 New Jersey Avenue NW,
10th floor
Washington, DC 20001
(800) 526-0859

You may ask for the research department, which may be able to give you information about issues that affect students, schools, and educational services (e.g., condom distribution, TB in schools). You can also try calling the local organization in your area.

American Forestry Association
P.O. Box 2000
Washington, DC 20013
(800) 368-5748

Information is available on the importance of forests to plants and animals.

American Foundation for AIDS Research (AmFAR)
733 Third Avenue, 12th floor
New York, NY 10017
(212) 682-7440

The staff can't answer questions over the phone, but they will send a booklet: Facts About AIDS and How Not to Get It.

American Health Care Association
1201 L Street NW
Washington, DC 20005-4014
(202) 842-4444

The public relations office will answer questions on health care issues. It has a resource center and library. They offer a teen award for the "Teen Volunteer of the Year."

American Humane Association
63 Inverness Drive E.
Englewood, CO 80112
(800) 227-4645
(303) 792-9900 (in Colorado)

Information on animal and child welfare is available.

American Institute for Cancer Research (AICR)
1759 R Street NW
Washington, DC 20069
(800) 843-8114
(202) 328-7744 (in Washington, DC)

The education and publication departments will answer questions and send free information on diet, nutrition, smoking, and cancer. Registered dieticians will answer your questions within 48 hours. If you send a letter, address it to "Nutrition Hotline."

American Medical Association (AMA)
515 North State Street
Chicago, IL 60610
(312) 464-4818

Ask for the library and the staff there will try to answer your questions on medicine and medical ethics.

American Red Cross
Program & Services Dept.
431 18th Street NW
Washington, DC 20006
(202) 728-6475

Also ask for information on the International Red Cross.

American Society for the Prevention of Cruelty to Animals (ASPCA) [National Headquarters]
Education Dept.
441 East 92nd Street
New York, NY 10128
(212) 876-7700

The education department will not answer questions over the phone, but you can request over the phone or by mail written information regarding animal testing and other animal welfare and protection issues.

ARC
P.O. Box 300649
Arlington, TX 76010
(800) 433-5255

ARC is a national organization dealing with mental retardation. Written requests for information are preferable. Information packets are available.

Bread for the World
802 Rhode Island Avenue NE
Washington, DC 20018
(202) 269-0200

This is a national advocacy group for rechanneling funds from other areas to be used for food assistance to the world's hungry and for developmental food programs in poor countries. The staff will answer questions over the phone or by mail. There are regional offices in Minneapolis, San Francisco, Chicago, and Denver.

CARE
660 First Avenue
New York, NY 10016
(212) 686-3110, ext. 341

CARE runs programs on health, nutrition, environmental protection, emergency aid, and help for small business in 39 developing countries.

Center for the Biology of Natural Systems (CBNS)
Queens College/CUNY
Flushing, NY 11367
(718) 670-4180

It conducts research and identifies solutions to problems involving alterna-

tive agriculture, environmental problems, and sustainable resource management.

Center for Science in the Public Interest (CSPI)
1875 Connecticut Avenue NW, Suite 300
Washington, DC 20009-5728
(202) 332-9110

Information is available on nutrition, food safety, organic agriculture, genetic engineering, and alcohol policies. It publishes a Nutrition Action Health Letter ten times a year. The staff prefers written requests, but if you call, call on a weekday between 3 P.M. and 4 P.M. (Eastern Time) and you will get a research person who will answer your questions or refer you elsewhere. There is no free literature.

Council for Agricultural Science and Technology
137 Lynn Avenue
Ames, IA 50010
(515) 292-2125

It has information on careers in agriculture and twice a year it publishes a science magazine for junior high and high school students.

Cystic Fibrosis Foundation
[National Headquarters]
6931 Arlington Road
Bethesda, MD 20841
(800) 334-4823

It provides both written and oral information.

David E. Luginbuht Institute
37 Franklin Street
Vernon, CT 06066
(203) 871-7599

It works on protecting endangered and threatened species and improving the coexistence of all living things on the earth through research.

Defenders of Wildlife
1244 19th Street NW
Washington, DC 20036
(202) 659-9510

It has a booklet titled Deadly Throwaways *about how litter harms animals. You can also get a copy of the endangered species list.*

Disabilities Rights Education and Defense Fund
2212 Sixth Street
Berkeley, CA 94710
(800) 466-4232

The small staff will try to answer questions over the phone, and written questions as well.

Earth Island Institute
300 Broadway, Suite 28
San Francisco, CA 94133
(415) 788-3666

This organization is made up of other environmental groups. It has information on endangered species and many other subjects.

Ecology Center
2530 San Pablo Avenue
Berkeley, CA 94702
(415) 548-2220

The staff will answer your questions or send you literature about environmental problems and solutions.

Environmental Action Coalition
652 Broadway
New York, NY 10012
(212) 677-1601

The staff will answer questions and distribute written materials. It focuses on urban issues.

Environmental Defense Fund (EDF)
[National Headquarters]
257 Park Avenue So.
New York, NY 10010
(212) 505-2100
(800) 225-5337

General environmental resources are available free by phone or mail. In addition, the EDF distributes a packet of information just for students on nine critical environmental issues.

Food and Nutrition Info Center
National Agricultural Library
Room 304
10301 Baltimore Boulevard
Beltsville, MD 20705
(301) 504-5719

This hotline is answered by registered dietitians. They also send written materials, or refer you to other organizations that can answer your questions.

Food Research & Action Center (FRAC)
1875 Connecticut Avenue, NW
Suite 540
Washington, DC 20009
(202) 986-2200

Ask for the communications department to answer questions.

Food & Water
225 Lafayette Street, Suite 612
New York, NY 10012
(212) 941-9340

It works on issues of food safety, particularly food irradiation.

Foundation for Biomedical Research
818 Connecticut Avenue, NW
Washington, DC 20006
(202) 457-0654

It deals only with animal research. Free information packets are available, but there is a charge for other literature, posters, tapes, and so on.

Future Farmers of America (FFA)
[National Headquarters]
5632 Mt. Vernon Memorial Highway
P.O. Box 15160
Alexandria, VA 22309-0160
(703) 360-3600

It has printed materials on agricultural careers, world agriscience studies, and food conservation.

Geothermal Education Office
664 Hilary Drive
Tiburon, CA 94920
(800) 866-4436

This organization serves students grades K-12. Leave your name, phone number, and address on the tape with your questions and the staff will send you materials. It also occassionally publishes a fact sheet called The Steam Press and have a booklet on geothermal energy. Educators can get even more information.

Greenpeace, USA
[National Headquarters]
1436 U Street NW
Washington, DC 20009
(202) 462-1177

When you call or write, ask for the local Greenpeace office nearest you (or check your phone book). Information is available on saving whales and other animals, as well as rain forests. Environmentalists in the information services division will answer your questions and send you free literature. The organization also publishes a newsletter.

Human Nutrition Information Service
USDA
6505 Belcrest Road
Hyattsville, MD 20782
(301) 436-7725 ext. 8617 (public affairs)

You can call or write for information on nutrition.

Human Nutrition Research Center on Aging
711 Wash Street
Boston, MA 02111
(617) 556-3000

It is a USDA organization affiliated with Tufts University. Ask for the Nutrition Library.

Humane Society of the U.S.
2100 L Street NW
Washington, DC 20037
(202) 452-1100

Information on the humane treatment of animals is available.

National Abortion Rights Action League (NARAL)
1101 14th Street NW, 5th floor
Washington, DC 20005
(202) 408-4600

The staff here only will answer written questions.

National Audubon Society, Inc.
[National Headquarters]
950 Third Avenue
New York, NY 10022
(212) 832-3200

Contact your local chapter to discuss environmental topics affecting Earth and its plants and animals. Check your local telephone book or call or write to the national headquarters.

National Center for Nutrition & Dietetics (NCND)
216 West Jackson Boulevard, Suite 800
Chicago, IL 60606-6995
(800) 366-1655 (consumer nutrition hotline)
(312) 899-4853 (in Illinois)

The center has a consumer nutrition hotline staffed by registered dietitians. The will send you free brochures about food, nutrition information, and food labeling.

National Coalition Against the Misuse of Pesticides
701 E Street SE, Suite 200
Washington, DC 20003
(202) 543-5450

This organization has two staff scientists to help answer your questions and send out information packets on the misuse of pesticides.

National Dairy Council
6300 North River Road
Rosemont, IL 60018-4233
(708) 803-2000

Free brochures are available from the council. For a fee, educational kits are available for all age groups, preschool through high school. Ask for the address and phone number of your state's regional office.

National Education Association (NEA)
NEA Health Information
1590 Adamson Parkway, Suite 260
Morrow, GA 30260
(404) 960-1325

Contact the above location to get information on health and education.

National Geographic Society
Dept. 01191
17th & M Streets NW
Washington, DC 20036
(800) 638-4077

Ask for the indexing division and it will do a free search (whether or not you are a member) for any information the society has on the subject you need. Junior membership in the society includes a subscription to the monthly magazine, National Geographic World.

National HIV and AIDS Info Service Hotline
(800) 342-AIDS

This is a telephone service open 24 hours a day, 7 days a week. It is sponsored by the Centers for Disease Control (CDC), and operated by the American Social Health Association. The staff will answer questions, send out information, and make referrals.

National Institute of Child Health & Human Development
Office of Research Reporting
9000 Rockville Pike
Building 31, Room 2832
Bethesda, MD 20894
(301) 496-5133

Call or write it for information or reference materials on children's health issues.

National Organic Farmers Association
P.O. Box 454
Ithaca, NY 14851
(607) 648-5557

The association is an educational group promoting organic and sustainable agriculture.

National Right to Life Committee
419 7th Street, Suite 500
Washington, DC 20004
(202) 626-8800

Ask for the Education Trust fund and the staff will send you a student packet.

National Wildlife Federation
1400 16th Street NW
Washington, DC 20036
(202) 797-6800

It has limited resources and does not have a staff large enough to spend time answering your questions or sending you materials. However, you may want to try writing anyway. There are two biologists on staff.

Nature Conservancy
[National Headquarters]
1815 North Lynn Street
Arlington, VA 22209
(703) 841-5300

It has biologists on staff and works to protect endangered plants and animals and promote biological diversity by acquiring land.

Oxfam America
115 Broadway
Boston, MA 02116
(617) 482-1211

or

Oxfam America
4797 Telegraph Avenue, Suite 201
Oakland, CA 94609
(510) 562-4388

Its education and outreach department answers questions and has bulletins and a free newsletter on hunger in America.

Population Institute
107 2nd Street, NE
Washington, DC 20002
(202) 544-3300

The education department or media coordinator can answer questions on Third World populations and international family planning issues.

Population Reference Bureau
1875 Connecticut Avenue NW
Washington, DC 20009-5728
(202) 483-1100

Inexpensive statistical charts and population information are available.

Public Interest Research Group (PIRG) [National Headquarters]
215 Pennsylvania Avenue SE
Washington, DC 20003
(202) 546-9707

There are offices in almost every state dealing with that state's local issues. This is a research and advocacy organization directed by college and university students with a professional staff in some areas.

Planned Parenthood of America
810 Seventh Avenue
New York, NY 10019
(212) 541-7800

It has information on birth control, abortion services, and sexually transmitted diseases. There are also state and local chapters. Check your telephone book.

Population Crisis Committee
1120 19th Street NW
Washington, DC 20036
(202) 659-1833

It is an educational and research organization.

Sierra Club [National Headquarters]
730 Polk Street
San Francisco, CA 94109
(415) 776-2211

The public information department answers questions over the phone from 8:30 A.M. to 1:00 P.M. (Pacific Time). It provides information on many environmental issues and distributes the Green Guide, *a list of 470 free or inexpensive sources of information on leading environmental issues.*

20-20 Vision [National Headquarters]
69 South Pleasant Street #203
Amherst, MA 01002
(413) 549-4555

This service keeps track of which government officials are involved with what issues.

United Nations Population Fund
220 E. 42nd Street
New York, NY 10017
(212) 297-5000

You have to be persistent to get through to them on the telephone.

World Health Organization (WHO) [United States Headquarters]
525 23rd Street NW
Washington, DC 20037
(202) 861-3200

Ask for the public information department. It publishes World Health *magazine which discusses health problems and how to prevent them.*

World Institute for Disabilities
510 16th Street
Oakland, CA 94612-1502
(510) 763-4100

Call or write for information.

World Wildlife Fund (WWF)
1250 24th Street NW
Washington, DC 20037
(202) 293-4800

It has information on rain forests and the protection of endangered wildlife, including elephants.

Zero Population Growth
1400 16th Street NW
Washington, DC 20036
(202) 332-2200

The staff answers questions and sends written information on world-wide population issues.

ORGANIZATIONS FOR YOUNG ADULTS

While the organizations mentioned above are for people of all ages, the following groups are targeted specifically for young adults.

Children Now
1930 14th Street
Santa Monica, CA 90404
(310) 399-7444

This is an advocacy group campaigning for better health care and services for children. It is a good source for referrals.

National Association for Humane Environmental Education
67 Salem Road
East Haddam, CT 06423
(203) 434-0172

This is a division of the Humane Society of the United States. It has information appropriate for students up to grade 12.

Kidsnet
6856 Eastern Avenue NW, Suite 208
Washington, DC 20012
(202) 291-1400

It has information geared to 6 to 12 year olds on AIDS, food labeling, and other things.

Mothers and Others for a Livable Planet
40 West 20th Street
New York, NY 10011
(212) 727-4474

Mothers and Others for a Livable Planet is a special project of the Natural Resources Defense Council (NRDC), (212) 727-4452. It concentrates on health and environmental hazards affecting children.

Renew America
1400 16th Street NW, Suite 710
Washington, DC 20036
(202) 232-2252

Ask for the children's program. The staff there collects true stories about children who make a difference and can refer you to other organizations and programs that can answer your specific questions.

Student Action Corps for Animals
P.O. Box 15588
Washington, DC 20003
(202) 543-8983

It encourages students to work to protect animals.

Teens Teaching AIDS Prevention
(Teens TAP)
(800) 234-8336

This is a hotline which is open seven days a week from 4:00 P.M. to 8:00 P.M. (Central Time).

PUBLICATIONS OF INTEREST

Here's a list of both books and magazines that can provide valuable information on the issues in this book. You may be able to find many of them at your local library.

Books

Carson, Rachel L., *Silent Spring,* NY: Ballantine Books, 1982.

Elkington, John, *Going Green: A Kids' Handbook to Saving the Planet,* NY: Viking, 1990.

Erickson, Judith, *Directory of American Youth Organization: A Guide to Over 400 Clubs, Groups, Troops, Teams, Societies, Lodges and More for Young People,* Minneapolis: Free Spirit Publishing, Inc.

Goldsmith, Edward; and Hildyard, Nicolas, *The Earth Report: The Essential Guide to Global Ecological Issues,* Price Stern Sloan, 1988.

Graham, *Nature Directory: A Guide to Environmental Organizations,* NY: Walker & Co., 1991.

Jarmul, David, *Headline News, Science Views,* Wash., DC: National Academy Press, 1991. 75 op-ed pieces from newspapers on science issues.

Javna, John and The Earthworks Group, *50 Simple Things KIDS Can Do to Save the Earth,* Berkeley, CA: Andrews & McMeel, 1990.

Lewis, Barbara A., *The Kids Guide to Social Action,* Minneapolis: Free Spirit Publishing Inc., 1991.

McKisson, Mikki; and Campbell, Linda McRae, *Our Only Earth* series, Tucson, AZ: Zephyr Press, 1990.

Nations, James D., *Tropical Rainforests,* NY: Franklin Watts, 1988.

Rifkin, Jeremy, *The Green Lifestyle Handbook: 1001 Ways You Can Heal the Earth,* NY: Henry Holt & Co., 1990.

Rossbacher, Lisa A., *Career Opportunities in Geology and the Earth Sciences,* NY: Arco, 1983.

If you want to learn still more about any of these module topics, you can find the following reference books at your local library. These books offer you lists of other books on these and related topics.

Malinowsky, *Best Science & Technology Reference Books for Young People,* Phoenix: Oryx Press, 1991.

Books for the Teenage, produced annually by the New York Public Library, NY.

Newspapers

In addition to their regular stories on science, many newspapers have a section once a week devoted to science and/or health. Below is a list of the largest newspapers that have science or health sections. Next to the name of the newspaper is the name of its science section and the day of the week the science section runs in the paper. Your local library probably subscribes to one or more of the newspapers on this list. Also, your local newspaper may have its own science and/or health section. Look in these newspapers to find articles on science and health topics that interest you.

Name	Section Name	Day Section Appears
Baltimore Sun	Health	Tuesday
Boston Globe	Health/Science	Monday
Boston Herald	Health	Monday
Cleveland Plain Dealer	Health & Science	Tuesday
Columbus Dispatch	Discovery	Sunday
Dallas Morning News	Discoveries	Monday
Detroit Free Press	Science & Medicine	Tuesday
Detroit News	Science	Thursday
Hartford Courant	Health & Science	Thursday
Houston Chronicle	Discovery	Monday
Los Angeles Times	Science/Medicine	Monday
Memphis Commercial Appeal	Mid-South Medicine	Sunday
Milwaukee Journal	Health	Monday
New York Newsday	Discovery	Tuesday
New York Times	Science Times	Tuesday
Newark Star Ledger	Health & Fitness	Sunday
Orange County Register (CA)	Health-Tech	Thursday
Philadelphia Daily News	Health	Wednesday
Portland Oregonian	Science	Thursday
San Diego Union	Quest	Monday
San Jose Mercury News	Science & Medicine	Tuesday
Seattle Times	Discovery	Monday
Washington Post	Health	Tuesday

Magazines

National Geographic World
Dept. 01191
17th & M Streets, NW
Washington, DC 20036
(800) 638-4077 (customer service)

This monthly magazine, published by the National Geographic Society, covers science, animals, nature, geography, and the environment. For grades 3 through 8, ages 8-13 years.

NatureScope
National Wildlife Federation
Correspondence Division
1400 16th Street NW
Washington, DC 20036
(800) 432-6564

NatureScope is published by the National Wildlife Federation. There are 18 separate issues on environmental topics, which are periodically updated. It is geared for students in kindergarten through eighth grade.

Scholastic Choices
c/o Scholastic, Inc.
2931 East McCarty Street
Jefferson City, MO 65102-9962
(800) 631-1586

This is a magazine on health and nutrition for grades 7-12. It is published 8 times a year from September to May.

Science World
c/o Scholastic, Inc.
2931 East McCarty Street
Jefferson City, MO 65102-9962
(800) 631-1586

This is a magazine on earth science, physical science, life science and health, space exploration and technology for grades 7-10. It is published 14 times a year from September to May.

Straight Talk
c/o Rodale Press
33 East Minor Street
Emmaus, PA 18098
(215) 967-8660

This is published 4 times during the school year for grades 7-12. Only health issues are covered including health and disease, substance abuse, AIDS, self-esteem, smoking and teen relationships.

3 2 1 Contact
P.O. Box 53051
Boulder, CO 80322-3051
(303) 447-9330

This children's magazine covers science topics. It's published 10 times a year (excluding February and August), by Children's Television Workshop for students 8 to 14 years old.

Other

Green Guide
c/o Sierra Club Environmental
   Education Committee
Dept. 5A
P.O. Box 7959
San Francisco, CA 94120
(415) 776-2211

A list of 470 free or inexpensive sources of information on environmental issues.

Superintendent of Documents
U.S. Government Printing Office (GPO)
Washington, DC 20402
(202) 783-3238

You can write to the GPO and ask them to send you a list of publications on any topic you're interested in. They publish over 30,000 books, booklets, pamphlets, and other documents. Many of their publications are free.

GOVERNMENT RESOURCES

Federal Agencies

Below is a list of relevant federal government offices and agencies. You should direct your questions to the public information office, unless otherwise indicated. The people there will either answer your question, or connect you to someone who can. If you don't have the number for a specific government office, call (202) 245-6999 for the federal government switchboard operator.

U.S. Bureau of the Census
Population Division
Rm. 2375
Federal Building 3
Suitland, MD 20746
(301) 763-5002 (general info line)

It has U.S. population information and some free written materials.

U.S. Dept. of Agriculture (USDA)
14th Street & Independence Avenue, SW
Washington, DC 20250
(202) 720-2791

It works to market farm products and combat hunger and malnutrition.

U.S. Dept. of Commerce
Office of the Under Secretary
National Oceanic and Atmospheric Administration
Room 5128
Washington, DC 20230
(202) 606-4380

It has information about the extinction of animals in general, including whales and dolphins.

U.S. Environmental Protection Agency (EPA)
Office of External Relations and Education
Youth Programs
401 M Street, SW
Washington, DC 20460
(202) 260-4454

It has environmental educational materials for young people. Write for a list of publications. If you have specific questions you will be directed to the proper office.

U.S. Fish and Wildlife Service (U.S. FWS)
Dept. of the Interior
18th and C Streets NW
Washington, DC 20240
(202) 208-5634

General information is available on endangered species. For more specific information, contact the Endangered Species Office of the USF & W Service, 4401 North Fairfax Drive, Arlington, VA 22203 (703) 358-2171. The staff there can tell you the current status of species on the federal list of threatened and endangered species.

U.S. Food and Drug Administration (FDA)
Freedom of Information Office
5600 Fishers Lane
Rockville, MD 20857
(301) 443-1544

The FDA will send free written information in response to written requests on topics such as food safety, additives and food treatments (such as irradiation), medical devices, and specific drugs. However, the FDA will not give out information over the telephone.

U.S. Dept. of Health and Human Services
200 Independence Avenue SW
Washington, DC 20201
(202) 245-6296

This agency deals with health issues and research, health-care policies, and human needs.

U.S. Department of Health and Human Services
Centers for Disease Control (CDC)
1600 Clifton Road, NE
Atlanta, GA 30333
(404) 639-3311

It publishes weekly and monthly reports on diseases. If you write to the CDC, be sure to state in the letter the specific disease you have a question about. The CDC sponsors the National HIV and AIDS Info Service Hotline *listed above under* **General Organizations.**

U.S. Dept. of Health and Human Services
National Center for Health Statistics
6525 Belcrest Road
Hyattsville, MD 20782
(301) 436-8500

As its name suggests, this agency is a resource for health statistics.

U.S. Dept. of Health and Human Services
National Institute of Environmental Health Sciences (NIEHS)
Public Affairs Office
P.O. Box 12233
Research Triangle Park, NC 27709
(919) 541-3345

It is the principal federal agency for biomedical research on the effects of environmental agents on the health of humans.

U.S. Dept. of Health and Human Services
National Institute of Health (NIH)
900 Rockville Pike
Bethesda, MD 20892
(301) 496-4000

Call the main number and you will be routed to the proper information office. The institute will send you materials that are appropriate for someone your age.

U.S. Dept. of the Interior
1800 C Street, NW
Washington, DC 20240
(202) 208-3100

This department is responsible for issues relating to national parks, forests, wildlife and other resources. However, call the U.S. Fish & Wildlife Service for information on threatened or endangered species.

U.S. Dept. of Justice
Immigration and Naturalization Service (INS)
425 I Street NW
Washington, DC 20536
(202) 514-4330

Immigration information officers can answer your questions on United States immigration laws, policies, and regulations.

U.S. Public Health Service
Public Affairs
Hubert H. Humphrey Building
200 Independence Avenue, SQ,
   Rm. 725-H
Washington, DC 20201
(202) 245-6768

The U.S. Public Health Service has information on AIDS, adoption, surrogacy, and many other health issues.

Government Officials

You can also call or write to key government officials to express your opinion or ask a question. For the correct address and/or name of your government representatives, check the blue pages in your phone book (the government section), go to your town hall, or ask your parent, teacher, or local librarian for help.

The President
The White House
Washington, DC 20500

The Vice President
The White House
Washington, DC 20500

The Honorable [name of member of president's cabinet]
The Secretary of [Agriculture, Health, Interior, etc.]
Washington, DC 20301

The Honorable [name of U.S. Senator]
United States Senate
Washington, DC 20510

The Honorable [name of U.S. Representative]
House of Representatives
Washington, DC 20515

The Honorable [name of state governor]
Governor of [state]
State Capital, Room #
[city, state, zip code]

The Honorable Mayor [name of mayor]
The Office of the Mayor
[street address]
[city, state, zip code]

In addition, you can call any congressional office and ask the staff there about the Congressional Research Service. Through the office, you can request a free background package on specific population, science, or environmental issues (e.g., the activities of the House Select Committee on Hunger).

State Fish and Wildlife Agencies

Here's a list of the mailing address, telephone number, and title of the person in charge at the main state fish and wildlife agency for each of the states in the United States and the District of Columbia. They generally protect, manage, and enhance fish and wildlife resources, and enforce the state's fish and game laws. Their staffs will usually answer your questions or send you written materials to help you learn more about these topics. You can also contact your city, county, or town government to find out if there's a local fish and wildlife agency or similar department. Check the blue pages of your telephone book.

ALABAMA
Commissioner
Dept. of Conservation & Natural Resources
64 North Union Street
Room 702
Montgomery, AL 36130
(205) 242-3486

ALASKA
Commissioner
Dept. of Fish & Game
P.O. Box 25526
Juneau, AK 99802-5526
(907) 465-4100

ARIZONA
Director
Dept. of Game & Fish
2222 West Greenway Road
Phoenix, AZ 85023
(602) 942-3000

ARKANSAS
Director
Game & Fish
 Commission
2 Natural Resources
 Drive
Little Rock, AR 72205
(501) 223-6305

CALIFORNIA
Director
Dept. of Fish & Game
1416 Ninth Street,
 12th floor
Sacramento, CA 95814
(916) 653-7664

COLORADO
Director
Div. of Wildlife
Dept. of Natural
 Resources
6060 Broadway
Denver, CO 80216
(303) 297-1192

CONNECTICUT
Director
Bureau of Fish &
 Wildlife
Dept. of Environmental
 Protection
165 Capitol Avenue
Room 255
Hartford, CT 06106
(203) 566-2287

DELAWARE
Director
Div. of Fish & Wildlife
Dept. of Natural
 Resources &
 Environmental
 Control
P.O. Box 1401
Dover, DE 19903
(302) 739-5295

FLORIDA
Executive Director
Florida Game &
 Freshwater Fish
 Commission
620 South Meridian
 Street
Tallahassee, FL 32399
(904) 488-2975

GEORGIA
Director
Game & Fish Div.
Dept. of Natural
 Resources
Suite 1362-E
205 Butler Street, SE
Atlanta, GA 30334
(404) 656-3523

HAWAII
Administrator
Div. of Forestry &
 Wildlife
Dept. of Land & Natural
 Resources
1151 Punchbowl Street
Honolulu, HI 96813
(808) 587-0166

or

Division Head
Div. of Aquatic
 Resources
Dept. of Land & Natural
 Resources
1151 Punchbowl Street
Honolulu, HI 96813
(808) 587-0100

IDAHO
Director
Dept. of Fish & Game
600 South Walnut
 Street
P.O. Box 25
Boise, ID 83707
(208) 334-5159

ILLINOIS
Director
Dept. of Conservation
Lincoln Towers Plaza
524 South Second
 Street
Springfield, IL 62701
(217) 782-6302

INDIANA
Director
Fish & Wildlife Div.
Dept. of Natural
 Resources
402 West Washington
 Street, Room W273
Indianapolis, IN 46204
(317) 232-4091

IOWA
Administrator
Fish & Wildlife Div.
Dept. of Natural
 Resources
Wallace State Office
 Building.
Des Moines, IA 50319
(515) 281-5918

KANSAS
Secretary
Dept. of Wildlife &
 Parks
Landon State Office
 Building
900 SW Jackson Street,
 Room 502
Topeka, KS 66612-1233
(913) 296-2281

KENTUCKY
Commissioner
Dept. of Fish & Wildlife
 Resources
Tourism Cabinet
#1 Game Farm Road
Frankfort, KY 40601
(502) 564-3400

LOUISIANA
Secretary
Dept. of Wildlife & Fisheries
P.O. Box 98000
Baton Rouge, LA 70898
(504) 765-2800

MAINE
Commissioner
Dept. of Inland Fisheries & Wildlife
State House Station #41
Augusta, ME 04333
(207) 289-3371

MARYLAND
Director
Tidewater Administration (Fish)
Dept. of Natural Resources
Tawes State Office Building
Annapolis, MD 21401
(301) 974-2926

or

Director
Wildlife Administration
Dept. of Natural Resources
Tawes State Office Building
Annapolis, MD 21401
(301) 974-3195

MASSACHUSETTS
Director
Div. of Fisheries & Wildlife
Dept. of Fisheries, Wildlife & Environmental Law Enforcement
Room 1902
100 Cambridge Street
Boston, MA 02202
(617) 727-3155

MICHIGAN
Director
Dept. of Natural Resources
Mason Building
P.O. Box 30028
Lansing, MI 48909
(517) 373-2329

MINNESOTA
Director
Div. of Fish & Wildlife
Dept. of Natural Resources
500 Lafayette Road
St. Paul, MN 55155
(612) 296-3344

MISSISSIPPI
Director
Dept. of Wildlife, Fisheries & Parks
P.O. Box 451
Jackson, MS 39205
(601) 362-9212

MISSOURI
Director
Dept. of Conservation
P.O. Box 180
Jefferson City, MO 65102-0180
(314) 751-4115

MONTANA
Administrator
Dept. of Fish, Wildlife & Parks
1420 East Sixth Avenue
Helena, MT 59620
(406) 444-3186

NEBRASKA
Director
Game & Parks Comm.
2200 North 33rd Street
P.O. Box 30370
Lincoln, NE 68503
(402) 464-0641

NEVADA
Director
Dept. of Wildlife
P.O. Box 10678
Reno, NV 89710
(702) 688-1500

NEW HAMPSHIRE
Executive Director
Fish & Game Dept.
2 Hazen Drive
Concord, NH 03301
(603) 271-3512

NEW JERSEY
Director
Div. of Fish, Game & Wildlife
Dept. of Environmental Protection & Energy
CN 400
Trenton, NJ 08625
(609) 292-9410

NEW MEXICO
Director
Game & Fish Dept.
Villagra Building
408 Galisteo Street
Santa Fe, NM 87503
(505) 827-7911

NEW YORK
Division Director
Div. of Fish & Wildlife
Dept. of Environmental
  Conservation
50 Wolf Road
Albany, NY 12233
(518) 457-5690

NORTH CAROLINA
Executive Director
Wildlife Resources
  Commission
Dept. of Natural
  Resources &
  Community
  Development
512 North Salisbury
  Street
Raleigh, NC 27604
(919) 733-3391

NORTH DAKOTA
Commissioner
Game & Fish Dept.
100 North Bismarck
  Expressway
Bismarck, ND 58501
(701) 221-6300

OHIO
Chief
Div. of Wildlife
Dept. of Natural
  Resources
Building G-3
1840 Belcher Drive
Columbus, OH 43224
(614) 265-6305

OKLAHOMA
Director
Dept. of Wildlife
  Conservation
1801 North Lincoln
  Boulevard
Oklahoma City, OK
  73152
(405) 521-3851

OREGON
Director
Dept. of Fish & Wildlife
2501 SW 1st Avenue
Portland, OR 97207
(503) 229-5406

PENNSYLVANIA
Executive Director
Fish & Boat Commission
3532 Walnut Street
Harrisburg, PA 17109
(717) 657-4518

or

Executive Director
Game Commission
2001 Elmerton Avenue
Harrisburg, PA 17110
(717) 787-3633

RHODE ISLAND
Chief
Div. of Fish & Wildlife
Dept. of Environmental
  Management
Oliver Stedman
  Government Center
4808 Tower Hill Road
Wakefield, RI 02879
(401) 789-3094

SOUTH CAROLINA
Executive Director
Wildlife & Marine
  Resources Dept.
P.O. Box 167
Columbia, SC 29202
(803) 734-4007

SOUTH DAKOTA
Secretary
Dept. of Game, Fish, &
  Parks
Foss Building
523 East Capitol
  Avenue
Pierre, SD 57501
(605) 773-3387

TENNESSEE
Executive Director
Wildlife Resources
  Agency
P.O. Box 40747
Nashville, TN 37204
(615) 781-6552

TEXAS
Executive Director
Parks & Wildlife Dept.
4200 Smith School
  Road
Austin, TX 78744
(512) 389-4800

UTAH
Director
Div. of Wildlife
  Resources
1596 W.N. Temple
Salt Lake City, UT
  84116
(801) 538-4700

VERMONT
Commissioner
Dept. of Fish & Wildlife
Agency of Natural
  Resources
103 South Main Street,
Waterbury, VT 05671
(802) 244-7331

VIRGINIA
Executive Director
Dept. of Game & Inland
  Fisheries
4010 West Broad Street
Richmond, VA 23230
(804) 367-9231

WASHINGTON
Director
Dept. of Wildlife
600 N. Capitol Way
M/S: GJ-11
Olympia, WA 98501
(206) 753-5710

or

Director
Dept. of Fisheries
P.O. Box 4313-5
Olympia, WA 98504
(206) 753-6623

WASHINGTON, D.C.
Administrator
Dept. of Consumer &
  Regulatory Affairs
DC Fisheries & Wildlife
2100 Martin Luther
  King Jr. Avenue, SE
Suite 203
Washington, DC 20020
(202) 404-1155

WEST VIRGINIA
Chief
Div. of Wildlife
  Resources
Dept. of Natural
  Resources
State Capitol Complex,
  Building 3
1900 Kanawha
  Boulevard
Charleston, WV 25305
(304) 348-2771

WISCONSIN
Director
Bureaus of Wildlife &
  Fisheries
  Management
Dept. of Natural
  Resources
P.O. Box 7921
Madison, WI 53707
(608) 266-1877

WYOMING
Director
Game & Fish
  Commission
5400 Bishop Boulevard
Cheyenne, WY 82006
(307) 777-4600

SCIENCE MUSEUMS AND ZOOS

Science museums and zoos are great sources for information. The scientists on staff are a wealth of knowledge. Many places have scientists or educators on staff who are willing to answer your questions, but in some places the staff is too small to be able to spend time on the phone answering your questions. Instead, write to them at the following addresses. This is only a sampling of the larger museums and zoos. Check your local library or the yellow pages of the telephone directory for the names and locations of others.

American Museum of Natural History
Central Park West at 79th Street
New York, NY 10024
(212) 769-5100

Arizona Sonora Desert Museum
2021 North Kinney Road
Tucson, AZ 85743
Attention: Librarian
(602) 883-1380

This is an outdoor facility. The librarian will try to answer your specific questions. But if you want general information, he or she will send you written materials.

Audubon Park & Zoological Garden
P.O. Box 4327
New Orleans, LA 70178
(504) 861-2537

Here you'll also find an aquarium.

Carnegie Science Center
Allegheny Square Annex
Pittsburgh, PA 15212-5363
(412) 237-1800

The staff at the Carnegie Science Center prefers questions by mail because of their limited resources.

Chicago Zoological Park (Brookfield Zoo)
8400 W. 31st Street
Brookfield, IL 60513
(312) 485-0263

Cincinnati Zoo
3400 Vine Street
Cincinnati, OH 45220
(513) 281-4700 or 4701

The zoo library staff will provide both oral and written information. For animal career information, contact the animal conservation department.

Cranbrook Institute of Science
500 Lone Pine Road
Box 801
Bloomfield Hills, MI 48303-0801
(313) 645-3230

This is the address for the administration department and the education department at the institute.

Dallas Museum of Natural History/Dallas Aquarium
P.O. Box 26193
Fair Park Station
Dallas, TX 75226
(214) 670-8460

The Exploratorium
3601 Lyon Street
San Francisco, CA 94123
(415) 563-7337

Explorers Hall
National Geographic Society
17th & M Streets, NW
Washington, DC 20036
(202) 857-7588

Explorers Hall is operated by the National Geographic Society and is located at its headquarters.

Fernbank Science Center
156 Heaton Park Drive NE
Atlanta, GA 30307
(404) 378-4311

Ask for the education department.

The Franklin Institute Science Museum & Planetarium
20th & The Benjamin Franklin Parkway
Philadelphia, PA 19103
(215) 448-1200

Grindstone Bluff Museum & Environmental Education Center
P.O. Box 7965
Shreveport, LA 71107
(318) 425-5646

Houston Museum of Natural Science
1 Hermann Circle Drive
Hermann Park
Houston, TX 77030
(713) 639-4629

Kansas City Zoological Gardens
Swope Park
Kansas City, MO 64132
(816) 333-7406

Their collection includes endangered and threatened species.

Lawrence Hall of Science
University of Berkeley Campus
Berkeley, CA 94720
(510) 642-5133 or (510) 642-4193

You can call or write. If you write, put the topic of your question on the envelope (e.g., health and disease, food additives) so the letter can be directed to the correct department.

Liberty Science Center and Hall of Technology
75 Montgomery Street
Jersey City, NJ 07304-4629
(201) 451-0006

The staff in the education department of this new science center will try to answer your questions.

Los Angeles Zoo
Griffith Park
5333 Zoo Drive
Los Angeles, CA 90027
Attention: Education Dept.
(213) 666-4650

The zoo won't answer questions over the phone, but you can call or write to get its brochure on how to do research about animals in the library.

Louisville Zoo
1100 Trevilian Way
Louisville, KY 40213-1559
(502) 459-2181

Ask for the education department.

Marineland, Inc.
First Coast Highway A1A
St. Augustine, FL 32086
(904) 471-1111

Miami Metro Zoo
12400 SW 152 Street
Miami, FL 33177
(305) 251-0401

Ask for the curator.

Museum of History and Science
727 W. Main Street
Louisville, KY 40202-2681
(502) 561-6100

Call or write with your questions.

Museum of Science
Science Park
Boston, MA 02114
(617) 723-3500

Museum of Science & Industry
57th Street & Lake Shore Drive
Chicago, IL 60637
(312) 684-1414

National Center for Atmospheric Research
P.O. Box 3000
1850 Table Mesa Drive
Boulder, CO 80307-3000
(303) 497-1000

New York Hall of Science/Hands-on Science Museum
47-01 111th Street
Corona, NY 11368
(718) 699-0675

New York Zoological Park (Bronx Zoo)
185th Street & Southern Boulevard
Bronx, NY 10460
(212) 367-1010

North Carolina Museum of Life and Science
433 Murray Avenue
Durham, NC 27704
(919) 220-5551

The staff will answer either written or oral questions.

Ohio's Center of Science and Industry (COSI)
280 E. Broad Street
Columbus, Ohio 43215
(614) 228-2674

Ask for visitor's services and you will be directed to the appropriate person. Or call the education department at (614) 228-2674 ext. 257.

Oregon Museum of Science and Industry
4015 SW Canyon Road
Portland, OR 97221-2797
(503) 222-2828

Ask for the public relations department or the secretary of the science department (503) 274-4552. Scientists are on staff to answer questions and/or send out information.

Pacific Science Center
200 2nd Avenue, N.
Seattle, WA 98109
(206) 443-2001

Philadelphia Zoological Garden
34th Street & Girard Avenue
Philadelphia, PA 19104
(215) 243-1100

Pittsburgh Aviary
Allegheny Commons West
Pittsburgh, PA 15212
Attention: Education Dept.
(412) 323-7233

The education department will gladly answer your questions about birds over the phone.

Riverbank Zoo
P.O. Box 1060
Columbia, SC 29202
(803) 256-4773

The education department will answer your questions, but doesn't have any written materials.

San Diego Zoo
P.O. 551
San Diego, CA 92112
(619) 231-1515

The Science Museum of Minnesota
30 E. 10th Street
St. Paul, MN 55101
(612) 221-9488

The Science Place
P.O. Box 151469
Dallas, TX 75315
(214) 428-5555

Ask for education or programs department.

Washington Park Zoo
4001 SW Canyon Road
Portland, OR 97221
Attention: Education
(503) 220-2781

The staff will try to answer questions, but no literature is available.

Woodland Park Zoo
5500 Phinney Avenue North
Seattle, WA 98103
(206) 684-4840

The staff will try to answer questions, but no literature is available.

CAREER AND TRADE ORGANIZATIONS

If you're interested in possibly pursuing a career in a field related to the ones discussed in this book, or you're still interested in more information on these issues, try contacting the following organizations. They consist of individuals and groups who specialize in related careers.

In addition, you may want to contact your local colleges and universities. Many schools across the country have research departments that specialize in these science areas and may be willing to share information with you, answer questions or have you visit and see what they do.

American College of Healthcare Executives
840 North Lake Shore Drive
Chicago, Illinois 60611
(312) 943-0544

American Dietetic Association
216 W. Jackson Boulevard, Suite 800
Chicago, IL 60606
(312) 899-0040

American Institute of Biological Sciences
Office of Career Service
730 11th Street, NW
Washington, DC 20001-4584
(202) 628-1500

American Society for Biochemistry and Molecular Biology
9650 Rockville Pike
Bethesda, MD 20814
(301) 530-7145

American Society for Horticultural Science
701 North Saint Asaph Street
Alexandria, VA 22314
(703) 836-4606

American Society for Microbiology
Office of Education and Professional Recognition
1913 I Street, NW
Washington, DC 20006
(202) 737-3600

American Society of Agronomy and Crop Science Society of America
677 South Segoe Road
Madison, WI 53711
(608) 273-8080

American Society of Zoologists
104 Sirius Circle
Thousand Oaks, CA 91360
(805) 492-3585

Association of University Programs in Health Administration
1911 Fort Myer Drive, Suite 503
Arlington, VA 22209
(703) 524-5500

Botanical Society of America
c/o Dr. Robert H. Essman
American Journal of Botany
Ohio State University
1735 Neil Avenue
Columbus, OH 43210-1293
(614) 292-1293

Consortium of Social Science Associations
1625 I Street NW, Suite 911
Washington, DC 20006
(202) 842-3525

Ecological Society of America (ESA)
Arizona State University
Center for Environmental Studies
Tempe, AZ 85287
(602) 965-3000

International organization. Distributes free booklet, Careers in Ecology *(including plants, animals, humans and the environment).*

Federation of Environmental Technologists (FET)
P.O. Box 185
Milwaukee, WI 53201
(414) 251-8163

Food and Agricultural Careers for Tomorrow
Purdue University
127 Agricultural Administration Building
West Lafayette, IN 47907
(317) 494-4600 (ask for FACT office)

Institute of Food Technologists
221 North LaSalle Street, Suite 300
Chicago, IL 60601
(312) 782-8424

National Association of Environmental Professionals (NAEP)
P.O. Box 15210
Alexandria, VA 22309-0210
(703) 660-2364

National Association of Social Workers
750 1st Street NE
Washington, DC 20002
(202) 408-8600

National Environmental Training Association (NETA)
8687 Via De Ventura, Suite 214
Scottsdale, AZ 85258
(602) 951-1440

National Organic Farmers Association
P.O. Box 454
Ithaca, NY 14851
(607) 648-5557

National Organization for Human Service Education
P.O. Box 6257
Fitchburg State College
Fitchburg, MA 01420
(508) 345-2151 (main college number)

Office of Higher Education Programs
U.S. Dept. of Agriculture
Administration Building
14th Street and Independence Avenue, SW
Washington, DC 20250
(202) 447-2791

Society for Range Management
1839 York Street
Denver, CO 80206
(303) 355-7070

Society of American Foresters
5400 Grosvenor Lane
Bethesda, MD 20814
(301) 897-8720

For more information about these and other associations and careers, look for the following books in the reference section at your local library:

Encyclopedia of Associations, Deborah M. Burek, editor, published annually by Gale Research, Inc., Detroit

Occupational Outlook Handbook, published annually by the U.S. Dept. of Labor, Bureau of Labor Statistics.

Acknowledgments

Problem Solving

"Campaign Will Seek Child Nutrition Labels" by Warren E. Leary. *The New York Times,* January 14, 1992. Copyright © 1991/1992 by The New York Times Company. Reprinted by permission.

Extinction of Living Things

1. "Foxes Hunted to Save Rare Bay Birds," *The San Francisco Chronicle,* May 13, 1991, p. A-15. © *San Francisco Chronicle.* Reprinted by permission.

2. "No Room to Roam" by Marla Cone, Times Staff Writer, *Los Angeles Times,* December 21, 1990, p. T-1 (Orange County Edition). Copyright 1990, Los Angeles Times. Reprinted by permission.

3. "Should We Downlist Our National Symbol?" by Carrie Casey. *American Forests,* November 1990, p. 24. Reprinted by permission of Carrie Casey from *American Forests.*

4. "Agency Sets Owl Acreage at 6.9 Million" by Roberta Ulrich. *The Oregonian,* January 10, 1992, p. A-1. Reprinted by permission of Roberta Ulrich and *The Oregonian.*

5. "Alternative Energy vs. the Rain Forest" by David L. Chandler. *The Boston Globe,* August 12, 1991, p. 33. Reprinted courtesy of *The Boston Globe.*

6. "Barnyard Rarities Get Their Day in Sun at Beltsville Show" by Doug Birch. *Baltimore Morning Sun,* September 14, 1991, p. 3-A. Reprinted by permission of *Baltimore Morning Sun.*

7. "Pessimism Is Growing on Saving Pandas From Extinction" by Sher WuDunn. *The New York Times,* June 11, 1991, p. C-4. Copyright © 1991/1992 by The New York Times Company. Reprinted by permission.

8. "Horns of a Dilemma," *Westchester Citizen Register,* January 10, 1992, p. A-11. Reprinted by permission of Associated Press.

9. "Closing in on Wild Bird Trade," *The Washington Post,* June 11, 1991, p. A-1. © 1991 *The Washington Post.* Reprinted by permission.

Human Populations

1. "Groups Unite to Point Out Hazards of Overpopulation," *The Los Angeles Times,* May 23, 1991, p. 20. Copyright 1991, Los Angeles Times. Reprinted by permission.

2. "Running Out of Room" by Patricia Orwen. *The Toronto Star,* May 26, 1991, p. B-1. Reprinted with permission—The Toronto Star Syndicate.

3. "Busting the Boom: Population Control Works," *WorldPaper,* April 1991, p. 11. © *The WorldPaper,* April 1991. Reprinted with permission.

4a. "Norplant Renews Debate Over Forced Contraception Reproductive Rights," *The Morning Call,* January 13, 1991, p. A-3. Copyright © 1991 by The New York Times Company. Reprinted by permission.

4b. "Birth Control or Woman Control? Norplant Not Meant for Coercion" by Ellen Goodman. *Charlotte Observer,* February 19, 1991, p. 13-A. © 1991, The Boston Globe Newspaper Co./Washington Post Writers Group. Reprinted with permission.

5. "Abortion Issue Divides Advocates for Disabled," *The New York Times,* July 4, 1991, p. A-11. Copyright © 1991/1992 by The New York Times Company. Reprinted by permission.

6. "U.S. Laws Under Attack—A Call for an Immigration Policy" by Ramon McLeod, *San Francisco Chronicle,* July 4, 1991, p. A-1. By Ramon McLeod, © *San Francisco Chronicle.* Reprinted by permission.

7. "An Investment in American Citizenship," *The Washington Post,* September 29, 1991, p. A-1. © 1991 *The Washington Post.* Reprinted with permission.

8. "Administration on Aging Announces the National Eldercare Campaign" by Joyce T. Berry. *Aging,* Winter 1991. Reprinted by permission of the U.S. Administration on Aging.

Human Health and Disease

1. "U.S. Reorganizes Nutrition Advice. Food Educators Win Battle to Depict 5 Basic Groups in a Pyramid Design" by Marian Burros. *The New York Times,* April 28, 1992, p. A-14. Copyright © 1992 by The New York Times Company. Reprinted by permission.

2. "Schools Relax TB Deadline," *Newsday,* October 3, 1991, p. 4. A Newsday article reprinted by permission. Newsday, Inc., copyright 1991.

3. "Doctors Group Urges Tough Laws on Smoking," *Los Angeles Times,* March 7, 1991, Metro Section, p. 1. Copyright 1991, Los Angeles Times. Reprinted by permission.

4. "Senate Panel Hears of Pesticide Harm Abroad," *The Los Angeles Times,* June 6, 1991, p. 4. Copyright 1991, Los Angeles Times. Reprinted by permission.

5. "Better Safe than Sorry?" by Susan Tifft, with reporting by Katherine L. Mihok/New York and James Willweth/Los Angeles. *Time,* January 21, 1991. Copyright 1991 The Time Inc. Magazine Company. Reprinted by permission.

6. "HIV Tests in the Health Profession" by Malcolm Gladwell, Washington Post Staff Writer, *The Washington Post,* September 11, 1991, p. A-21. © 1991 THE WASHINGTON POST, reprinted with permission.

7a. "Behind Animal Testing Is Profit Motive" by Valerie Gaston, *Roanoke Times and World News,* October 18, 1991, p. A-10. © Valerie Gaston. Reprinted by permission.

7b. "Think Animal Testing Is Inhumane? Explain It to the Sick, Dying" by Dr. Kurt J. Isselbacher. *Charlotte Observer,* June 12, 1991, p. 15-A. Reprinted by permission of Kurt J. Isselbacher, M.D.

8. "Laying Siege to a Deadly Gene," *Time,* February 24, 1992, p. 60. Copyright 1992 The Time Inc. Magazine Company. Reprinted by permission.

9. "Group Legislations Target Women's Health Research" by Michael Clements. *The Detroit News,* April 8, 1991. Reprinted with permission of *The Detroit News,* a Gannett Newspaper, copyright 1991.

World Food Resources

1a. "Famine in Africa, the Other Desert Crisis," *The Arizona Republic,* March 7, 1991, p. A-18. Reprinted by permission of the Arizona Republic.

1b. "World Hunger Is Persistent but Not Inevitable" by George Moffett. *The Christian Science Monitor,* October 17, 1991, p. 6. Reprinted by permission from *The Christian Science Monitor,* © 1991 The Christian Science Publishing Society. All rights reserved.

1c. "In a Changing World, Little Has Changed for the Hungry," *St. Petersburg Times,* October 15, 1991, p. 11-A. Reprinted by permission of the St. Petersburg Times.

2a. "Hunger Said to Afflict 1 in 8 American Children," *The Washington Post,* March 27, 1991, p. A-4. © *The Washington Post.* Reprinted with permission.

2b. "Survey: 160,000 Georgia Children Go Hungry," *The Atlanta Journal,* March 27, 1991, p. C-3. Reprinted with permission from *The Atlanta Journal* and *The Atlanta Constitution.* Reproduction does not imply endorsement.

3a. "Sustainable Agriculture More Pragmatic Than Organic Farming," *St. Paul Pioneer Press,* January 17, 1991, p. 3-B. Reprinted by permission of Associated Press.

3b. "Organic Farming is the Solution to Residue Pesticides, Expert Says" by Reggie McLeod. *St. Paul Pioneer Press,* February 25, 1991, p. 13-A. Reprinted by permission of *St. Paul Pioneer Press.*

4. "State's Growth Threatens Way of Life in Rice Towns," *The Los Angeles Times,* April 7, 1991, p. A-1. Copyright, 1991, Los Angeles Times. Reprinted by permission.

5. "Irradiated Food Coming, But Not Without Protest," *The New York Times,* January 21, 1992, p. A-1. Copyright © 1991/1992 by The New York Times Company. Reprinted by permission.

6. "Genetic Diversity Prevents Blight, Spread of Famine," *The Star Ledger,* June 10, 1991. Reprinted by permission of The Star Ledger.

7. "Scientists Use Gene Alterations to Make Crops Resistant to Infestation," *Los Angeles Times,* October 14, 1991, Metro Section, p. B-3. Copyright 1991, Judy Berlfein. Reprinted by permission.

Art and Photos

Problem Solving **x–1:** Larry Lawfer/Picture Cube; **4–5:** Skjold Photography; **8:** Bob Daemmrich/Stock Boston; **14:** © 1991 Ziggy and Friends Inc./Distributed by Universal Press Syndicate; **18:** Bob Daemmrich/Stock Boston; **20:** © 1991 Ziggy and Friends Inc./Distributed by Universal Press Syndicate; **22:** Courtesy of the Atrium Society; **23.** Bob Daemmrich/Stock Boston.

Extinction of Living Things **24–25:** Michael Fugden/Earth Scenes; **27:** Mickey Gibson/Earth Scenes; **29:** Anthony Bannister/Animals, Animals; **30:** Leonard Lee Rue III/Animals, Animals; **31:** Mickey Gibson/Earth Scenes; **33:** Mickey Gibson/Earth Scenes; **35:** Grant Heilman; **37:** Len Rue Jr./Animals, Animals; **38:** Ken Gardner; **43:** Adapted from Dennis Lowe/Los Angeles Times. Copyright 1990, Los Angeles Times. Adapted by permission; **44:** Gary Ambrose/Los Angeles Times. Copyright 1990, Los Angeles Times. Reprinted by permission; **45:** Gary Ambrose/Los Angeles Times. Copyright 1990, Los Angeles Times. Reprinted by permission; **49:** © Didier J. Lindsey; **53:** Courtesy of The Oregonian; **54:** Courtesy of The Oregonian; **57:** Reprinted by permission of Universal Press Syndicate; **58:** J.D. Griggs/U.S. Geological Survey; **59:** Reprinted by permission of Universal Press Syndicate; **67:** © George B. Schaller; **68:** Animals, Animals; **71:** Adapted by permission of Westchester Citizen Register; **74:** HSUS/Dantzler; **75:** Humane Society of the U.S./Dantzler © 1991/HSUS; **78:** Drawing by W. Miller, © 1991 The New Yorker Magazine, Inc; **79:** New York Zoological Society

Human Populations **80–81:** Susan Leavines/Photo Researchers; **83:** Bernard Pierre Wolff/Photo Researchers; **84:** Fay Torresyap/Stock Boston; **85:** Peter Menzel/Stock Boston; **87:** Jan Lukes/Photo Researchers; **89:** Chester Higgens/Photo Researchers; **97:** Adapted with permission—The Toronto Star Syndicate; **103:** © Sidney Harris; **107:** Courtesy Wyeth-Ayerst Laboratories; **117:** © San Francisco Chronicle. Adapted by permission; **121:** Courtesy of Senator Dale Bumpers; **127:** Diane Gerard.

Human Health and Disease **128–129:** Lean Claude Lejune/Stock Boston; **131 (left):** Steven Stone/Picture Cube; **131 (right):** J.D. Sloan/Picture Cube; **134:** Barbara Rios/Photo Researchers; **135:** Doug Plummer/Photo Researchers; **136:** Len Nelson; **142:** Copyright © 1992 by The New York Times Company. Adapted by permission; **150:** The Far Side Gallery by Gary Larson is reprinted by permission of Chronicle Features, San Francisco, CA; **154:** Courtesy of Senator Patrick J. Leahy; **167:** © 1971 by Sidney Harris—What's So Funny About Science (William Kaufmann, Inc); **170:** Copyright 1992 The Time Inc. Magazine Company. Adapted by permission; **177:** Courtesy of Just Say No International.

World Food Resources **178–179:** Ray Witlin/United Nations; **181:** Steve Hanson/Stock Boston; **183:** United Nations; **184:** Peter Vandermark/Stock Boston; **185:** Diane Rawson/Photo Researchers; **186:** Lynn Lemon/Photo Researchers; **191:** Esaias Baitel/Gamma Liaison; **193:** Betty Press; **194:** Wide World Photos; **199:** ©1991 Food Research and Action Center; **200:** Adapted with permission from the Atlanta Journal and The Atlanta Constitution. Reproduction does not imply endorsement; **201:** Courtesy the Center for Food Action; **202:** Reprinted by permission of Universal Press Syndicate; **210:** Copyright, 1991, Los Angeles Times. Adapted by permission; **215:** Adapted by permission of Associated Press; **216:** Copyright © 1992 by The New York Times Company. Adapted by permission; **218:** Copyright © 1992 by The New York Times Company. Adapted by permission; **226:** Copyright, 1991, Los Angeles Times. Adapted by permission; **229:** Spencer Grant/Stock Boston.

Skills **231:** Picture Cube; **232:** Rhoda Sidney; **234:** Skjold Photography; **237:** Picture Cube; **241:** Stock Boston; **243:** Stock Boston.

Glossary

acid rain: rain containing pollution, 31

action plan: identifies an action that can be done in order to get a solution accepted by others, 21

AIDS: acquired immune deficiency syndrome, a disease of the immune system, 133

baby boom: increased birth rate in the United States during the late 1940s and 1950s, 88

balanced diet: eating the proper amount of foods to be healthy, 134

belief: idea that a person holds to be true, whether it is really true or not, 15

benefit: positive, or good, result, 21

biosphere: thin zone of the earth that supports all life, 28

calorie: unit used to measure energy from foods, 185

carcinogen: substance that can cause cancer, 133

contagious: able to be transmitted from one person to another, 132

cost: negative, or bad, result, 21

crop yield: amount of a crop produced in a harvest, 182

deficiency disease: disease caused by having too little of something, 134

deforestation: clearing the forests of trees, 183

drug: chemical substance that causes a change in the mind or body, 136

ecosystem: living and nonliving things in an environment, together with their interactions, 28

endangered species: living things that are in danger of dying out, 32

engineer: person who plans, designs, and helps build complex devices, 8

evolution: process by which species of organisms change over time, 29

extinct: species that is no longer found alive, 27

famine: extreme shortage of food, 184

fossil fuels: fuels such as oil and natural gas that are developed from organic materials, 182

gene diversity: variety in the types of genes that living things have, 183

genetic diversity: variety in the traits of living things, 29

genetic engineering: methods used to change an organism's DNA, 169, 187

Green Revolution: rapid increase in the amount of food produced worldwide in the 1960s, 84, 187

habitat: place where an organism lives, 32

herbicide: chemical used to kill unwanted plants, 31, 182

HIV: human immunodeficiency virus, the virus that causes AIDS, 133

immigration: movement of people into an area or a country, 88

infectious disease: disease that can be passed from one person to another, 132

irradiate: to expose to radiation, 213

issue: question to be resolved, 14

malnutrition: condition that results when the body is not properly nourished, 185

mental health: the state of well-being of the mind, 131

noninfectious disease: disease that can not be passed from one person to another, 133

nutrient: material needed for growth, energy, and life processes, 134

organic compound: compound containing carbon found in things that are living or were once living, 28

organic farming: farming without the use of chemical fertilizers, pesticides, or herbicides, 187

pathogen: microscopic organism that causes disease, 132

persuasion: trying to convince people of a need for change, 21

pesticide: chemical used to kill unwanted insects, 182

physical health: the state of well-being of the body, 131

players: people who are involved in an issue, 14

position: how a person feels about an issue, 15

poverty cycle: process whereby the children of hungry and poor people grow up to be hungry and poor themselves, 186

science: human activity directed toward trying to understand nature, 7

scientific literacy: ability to use science and technology to solve everyday problems and make decisions about issues in our society, 10

side effect: an effect in addition to one that is desired, 21

social health: the state of well-being in relationships people have with one another, 131

society: a group made up of all individuals who are dependent upon one another, 9

starvation: state of hunger to the point of death, 190

subsistence agriculture: producing just enough food to survive, 183

sustainable farming: farming with few chemical fertilizers, pesticides, and herbicides, 187

technology: the application of knowledge, tools, and skills to make and do things that are useful in life, 7

toxin: poison, 132

trade-off: compromise that is arrived at by weighing the benefits of a course of action against the costs, 21

value: worth, or importance, a person places on something, 15

wellness: a condition of physical, mental, and social well-being, 131

wildlife refuge: place where living things are protected from being harmed by humans, 33

world grain surplus: amount of extra grain stored away worldwide, 187

Index

A

Abortion issues and disabled persons
 description of, 111–14
 exercises for studying, 114
Accidents, 132
Acid rain, 31
Administration on Aging (AOA), 123, 124–25
Africa, famine in, 190–92
Aging of America
 description of, 123–25
 exercises for studying, 125
AIDS (Acquired Immune Deficiency Syndrome)
 condom distribution in high schools, 156–59
 health care workers tested for, 160–63
 as a health issue, 133
Alcohol, wellness and, 137
Americans with Disabilities Act, 112–13
Animals used for medical research and testing
 description of, 164–68
 exercises for studying, 168
Awareness fair, how to set up an, 243

B

Baby boom, 88
Balanced diet, 134
Being objective, 14
Beliefs, 15
Benefits, 21
Berlfein, Judy, 225
Berry, Joyce T., 124
Biosphere, 28
Birch, Doug, 62
Birds
 foxes hunted to save clapper rail, 36–39, 42
 restricting trade of rare wild, 73–77

Birth control, condom distribution in high schools
 description of, 156–59
 exercises for studying, 159
Birth control versus reproductive rights, forced
 description of, 105–10
 exercises for studying, 110
Birth rate(s)
 limiting, 86–87
 versus death rate, 84–85
Brainstorming, 231
Bread for the World, 191, 192, 193, 195
Bumpers, Dale, 120
Burros, Marian, 141

C

California, rice production in
 description of, 207–12
 exercises for studying, 212
Canada, population problems and
 description of, 95–99
 exercises for studying, 99
Cancer, 133
Carcinogens, 133
Card catalog, how to use the, 238
Carlson, James A., 204
Casey, Carrie, 48
Cereals, using science and technology to study
 determining the best deal, 4–6
 using a spreadsheet, 6
Chandler, David L., 57
Children's Food Labeling Initiative, 11
China, limiting births in, 86
Cigarette vending machines, banning
 description of, 148–51
 exercises for studying, 151
Circle of Poison Prevention Act, 154
Clements, Michael, 174

Cone, Marla, 41
Condom distribution in high schools
   description of, 156–59
   exercises for studying, 159
Consumer advocacy group, 12
Convention on International Trade in Endangered Species (CITES) 32
Cougars, 41, 44, 45–46
Crop yields, 182
Cystic fibrosis, genetic engineering and
   description of, 169–72
   exercises for studying, 172

D

Data table, how to make a, 236–37
Death rate, birth rate versus, 84–85
DDT, 48
Deficiency diseases, 134
Deforestation, 183
Development versus extinction
   description of, 40–46
   exercises for studying, 46
Dinosaurs, 27
Disabled persons and abortion issues
   description of, 111–14
   exercises for studying, 114
Diseases
   AIDS, 133
   cystic fibrosis, 169–72
   deficiency, 134
   infectious, 132–33
   noninfectious, 133
   tuberculosis, 145–47
Downey, Thomas J., 199
Drugs, wellness and, 136–37

E

Eagles, extinction and bald
   description of, 47–51
   exercises for studying, 51
Ecosystems, 28
Edison, Thomas, 9
Endangered species, 32
Endangered Species Act, 32, 48, 49
Environment
   health problems caused by problems with the, 137
   population problems and impact on, 88
Evolution, 29
Exercise, wellness and, 135
Extinction
   approaches to saving species from, 32–33
   cougars, 41, 44, 45–46
   development versus, 40–46
   dinosaurs, 27
   discovery exercises and, 34–35
   eagles, 47–51
   effects of, 28–29
   exercises for studying, 39, 46, 51, 55, 60, 63, 69, 72, 77–79
   farm animals, 61–63
   foxes hunted to save clapper rails, 36–39, 42
   habitat loss and, 32
   human causes of, 30–31
   natural resources and, 29–30
   owls, spotted, 52–55
   pandas, 64–69
   pollution and, 31–32
   rain forest versus alternative energy, 56–60
   rate of, 27
   rhinoceroses, 70–72
   wild birds, 73–77

F

Famine
   in Africa, 190–92
   definition of, 184
   growth of, 195–96
   not inevitable, 192–94
Farm animals, preserving rare
   description of, 61–63
   exercises for studying, 63
Farming
   organic, 187, 205–6
   subsistence, 183–84
   sustainable, 187, 204–5
Farming techniques, changes in, 187
Food irradiation
   description of, 213–19

exercises for studying, 219
Food pyramid, 141–43
Food resources, problems and issues
  changes in farming techniques, 187
  crop yields, 182
  deforestation, 183
  description of, 183–87
  discovery exercises and, 188–89
  exercises for studying, 196, 202, 206, 212, 219, 223, 227–29
  famine, 184, 190–202
  food irradiation, 213–19
  future and, 187
  genetic diversity and, 220–23
  genetic engineering and, 187, 224–27
  health problems, 185
  high-yield crops, 182–83
  hunger and American children, 197–202
  malnutrition and undernutrition, 185–86
  organic farming, 187, 205–6
  pesticides and herbicides used, 182
  rice production in California threatened, 207–12
  shortages of food, 186–87
  soil erosion, 183
  sources for food, 182–83
  subsistence farming, 183–84
  sustainable farming, 187, 203–5
  water rights, 182
Foxes hunted to save clapper rails
  description of, 36–39, 42
  exercises for studying, 39

G

Gaston, Valerie W., 165
Genetic diversity
  description of, 29–30, 183, 220–23
  exercises for studying, 223
Genetic engineering and cystic fibrosis
  description of, 169–72
  exercises for studying, 172
Genetic engineering and farming
  description of, 187, 224–27
  exercises for studying, 227

Geothermal development, rain forest destruction versus, 56–60
Gladwell, Malcolm, 161
Goodman, Ellen, 109
Gramm, Phil, 120
Green Revolution, 84, 187

H

Habitat loss, extinction and, 32
Hacker, Holly K., 93
Health care workers tested for AIDS
  description of, 160–63
  exercises for studying, 163
Health issues
  accidents, 132
  AIDS, 133
  alcohol, 137
  animals used for medical research and testing, 164–68
  banning of cigarette vending machines, 148–51
  condom distribution in high schools, 156–59
  description of, 131–37
  discovery exercises and, 138–39
  drugs, 136–37
  environmental problems and, 137
  exercise and, 135
  exercises for studying, 144, 147, 151, 155, 159, 163, 168, 172, 175–77
  genetic engineering and cystic fibrosis, 169–72
  health care workers tested for AIDS, 160–63
  industry pressure versus public health, 140–44
  infectious diseases, 132–33
  lifestyle choices and, 136–37
  malnutrition and undernutrition, 185–86
  mental health, 131
  noninfectious diseases, 133
  nutrition and, 134–35
  pesticide used overseas, 152–55
  physical health, 131
  social health, 131
  stress, 135–36

tobacco, 136, 148–51
tuberculosis, 145–47
wellness, 131, 133–36
women and medical research and testing, 173–75
Helms, Jesse, 161
Herbicides, 31, 182
Holmes, Steven A., 112
Hunger and American Children. See also Famine
description of, 197–202
exercises for studying, 202

I

Immigration policy, population problems and U.S.
buying of American citizenship, 120–22
description of, 115–22
exercises for studying, 118, 122
need for the U.S. to have a policy, 116–18
Industry pressure versus public health
description of, 140–44
exercises for studying, 144
Infectious diseases, 132–33
Information
evaluating, 19
gathering, 17–19
methods for gathering, 19
skills needed for gathering, 231–38
sources of, 17–19
writing a letter to request, 234
Irradiation of food
description of, 213–19
exercises for studying, 219
Isselbacher, Kurt J., 166

J

Jackson, Robert L., 153
Journal, keeping a, 233

K

Kamen, Al, 120
Kennedy, Edward, 120

Kennedy, Robert, 117
Kidsnet, 11–12

L

Lancaster, John, 74
Leahy, Patrick J., 153–54
Leary, Warren E., 11
Letter, how to write a
to a member of Congress, 242
to request information, 234
Lewin, Tamar, 106
Lifestyle choices, wellness and, 136–37

M

McCoy, Kevin, 146
McLeod, Ramon G., 116
McLeod, Reggie, 205
MacPherson, Kitta, 221
Malnutrition and undernutrition, 185–86
Marshall, Alex, 101
Medical research and testing, animals used for
description of, 164–68
exercises for studying, 168
Medical research and testing, women and
description of, 173–75
exercises for studying, 175
Mental health, 131
Mihok, Katherine L., 157
Models, creating, 7
Moffett, George D., III, 192

N

News release, how to write a, 239
Noninfectious diseases, 133
Norplant implants, reproductive rights versus
description of, 105–10
exercises for studying, 110
Nutrients, 134
Nutrition
malnutrition and undernutrition,

185–86
wellness and, 134–35
Nutrition labels, studying
analyzing the issue, 13–16
description of, 4–6, 11–12
gathering information, 17–19
making a decision, 20–21
taking action, 21–22

O

Oakar, Mary Rose, 174
Observing nature, 7
Organic farming
definition of, 187
description of, 205–6
exercises for studying, 206
Orwen, Patricia, 96
Overpopulation
Canada and, 95–99
groups united to point out hazards of overpopulation, 92–94
Owls, extinction and spotted
description of, 52–55
exercises for studying, 55

P

Pandas, extinction and
description of, 64–69
exercises for studying, 69
Pathogens, 132
Patterns, seeking, 7
PCB poisoning, 32
Perl, Rebecca, 200
Persuasion, 21–22
Persuasive speech, how to make a, 240–41
Pesticide used overseas
description of, 152–55
exercises for studying, 155
Pesticides, impact of, 182
Physical health, 131
Pollution, extinction and, 31–32
Population density, 85
Population problems and issues
abortion issues and disabled persons, 111–14
aging of America, 123–25
birth rate versus death rate, 84–85
Canada and, 95–99
description of, 83–89
discovery exercises and, 90–91
exercises for studying, 94, 99, 104, 110, 114, 118, 122, 125–27
forced birth control versus reproductive rights, 105–10
groups united to point out hazards of overpopulation, 92–94
growth rates, 85–86
impact of technology on, 89
impact on the environment, 88
importance of population control, 100–104
limiting birth rates, 86–87
reasons for the increase in population, 84
in the U.S., 88–89
U.S. immigration policy, 115–22
Power plants, rain forest destruction versus, 56–60
Purvis, Andrew, 170

R

Rain forest versus alternative energy, 56–60
Records, information and, 18–19
Reproductive rights, birth control/Norplant implants versus
description of, 105–10
exercises for studying, 110
Rhinoceroses, dehorning
description of, 70–72
exercises for studying, 72
Rice production in California threatened
description of, 207–12
exercises for studying, 212
Rich, Spencer, 198
Roderick, Kevin, 208
Rohter, Larry, 214

S

Schroeder, Patricia, 174

Science
   definition of, 7
   exercises for studying, 13–23
   how technology and society interact with, 10
   how technology interacts with, 9
   methods of, 7
   processes of, 7
Scientific literacy, 10
Scientific method, 7
Scientist, who can be a, 7
Scrapbook, making a, 232
Shah, Reena, 195
Side effects, 21
Singapore, limiting birth in, 86
Social health, 131
Society
   definition of, 9
   exercises for studying, 13–23
   how technology and science interact with, 10
Soil erosion, 183
Spreadsheet, definition of, 6
Starvation, definition of, 190
Stress, 135–36
Subsistence farming, 183–84
Sustainable farming
   definition of, 187
   description of, 204–5
   exercises for studying, 206

T

Technologist, who can be a, 8–9
Technology
   definitions of, 7–8
   exercises for studying, 13–23
   how science and society interact with, 10
   how science interacts with, 9
   population problems and, 89
Telephone, gathering information with a, 235
Tifft, Susan, 157
Timber industry vs. spotted owls, 52–55
Tobacco
   banning of cigarette vending machines, 148–51
   wellness and, 136
Toxins, 132
Trade-offs, 21
Tuberculosis screening delayed
   description of issue, 145–47
   exercises for studying, 147

U

Ulrich, Roberta, 53
Undernutrition, 185–186
United States
   aging of the, 123–25
   hunger and children in the, 197–202
   immigration policy in the, 115–22
   population problems in the, 88–89
   rice production in California threatened, 207–12
U.S. Department of Agriculture (USDA), 140, 141, 143

V

Values, 15
Viruses 132–33

W

Water rights, food resources and, 182
Wellness
   definition of, 131
   maintaining, 133–36
Wielawski, Irene, 149
Wildermuth, John, 37
Wildlife corridors, 42–44
Wildlife refuges, 33
Willwerth, James, 157
Women and medical research and testing
   description of, 173–75
   exercises for studying, 175
Women's Health Equity Act, 174
World Conference on Environment and Development (WCED), 103–4
World grain supplies, 187
WuDunn, Sheryl, 65